Tibor Radó

On the Problem
of Plateau

Subharmonic Functions

Springer-Verlag Berlin Heidelberg GmbH

AMS Subject Classifications (1970): 49F10, 31C05

ISBN 978-3-642-98307-8   ISBN 978-3-642-99118-9 (eBook)
DOI 10.1007/978-3-642-99118-9

Library of Congress Catalog Card Number 71-160175.

Herstellung: Strauss & Cramer, Leutershausen

ERGEBNISSE DER MATHEMATIK
UND IHRER GRENZGEBIETE

HERAUSGEGEBEN VON DER SCHRIFTLEITUNG
DES
„ZENTRALBLATT FÜR MATHEMATIK"
ZWEITER BAND

————— 2 —————

ON THE PROBLEM
OF PLATEAU

BY

TIBOR RADÓ

WITH 1 FIGURE

Springer-Verlag Berlin Heidelberg GmbH

Contents.

Introduction.

The most immediate one-dimensional variation problem is certainly the problem of determining an arc of curve, bounded by two given points and having a smallest possible length. The problem of determining and investigating a surface with given boundary and with a smallest possible area might then be considered as the most immediate two-dimensional variation problem.

The classical work, concerned with the latter problem, is summed up in a beautiful and enthusiastic manner in DARBOUX's Théorie générale des surfaces, vol. I, and in the first volume of the collected papers of H. A. SCHWARZ. The purpose of the present report is to give a picture of the progress achieved in this problem during the period beginning with the Thesis of LEBESGUE (1902).

Our problem has always been considered as the outstanding example for the application of Analysis and Geometry to each other, and the recent work in the problem will certainly strengthen this opinion. It seems, in particular, that this recent work will be a source of inspiration to the Analyst interested in Calculus of Variations and to the Geometer interested in the theory of the area and in the theory of the conformal maps of general surfaces. These aspects of the subject will be especially emphasized in this report.

The report consists of six Chapters. The first three Chapters are concerned with investigations which yielded either important tools or important ideas for the proofs of the existence theorems reviewed in the last three Chapters.

Chapter I.

Curves and surfaces.

I.1. If $x = x(t)$, $y = y(t)$, $z = z(t)$, $a \leq t \leq b$ are the equations of a curve C, then under the usual classroom assumptions the length $l(C)$ of C is given by the formula

$$l(C) = \int_a^b \left[\left(\frac{dx}{dt}\right)^2 + \left(\frac{dy}{dt}\right)^2 + \left(\frac{dz}{dt}\right)^2 \right]^{\frac{1}{2}} dt. \tag{1.1}$$

If C reduces to a straight segment of length l, then the formula (1.1) reduces to $l = (l_1^2 + l_2^2 + l_3^2)^{\frac{1}{2}}$, where l_1, l_2, l_3 denote the lengths of the orthogonal projections of the segment upon the axes x, y, z (the coordinate system will always be supposed to be rectangular). The formula (1.1) is equally evident geometrically if C is a polygon. It is then clear that for a general curve C the formula results by approximating C by polygons[1]. As a matter of fact, (1.1) follows immediately by approximating C by an inscribed polygon.

I.2. If $x = x(u, v)$, $y = y(u, v)$, $z = z(u, v)$, (u, v) in some region R, are the equations of a surface S, then under the usual classroom assumptions the area $\mathfrak{A}(S)$ of S is given by the formula

$$\mathfrak{A}(S) = \iint_R \left[\left(\frac{\partial(y, z)}{\partial(u, v)}\right)^2 + \left(\frac{\partial(z, x)}{\partial(u, v)}\right)^2 + \left(\frac{\partial(x, y)}{\partial(u, v)}\right)^2 \right]^{\frac{1}{2}} du \, dv. \tag{1.2}$$

If S reduces to a triangle with area Δ, then (1.2) reduces to

$$\Delta = (\Delta_1^2 + \Delta_2^2 + \Delta_3^2)^{\frac{1}{2}},$$

where $\Delta_1, \Delta_2, \Delta_3$ denote the areas of the triangles obtained by orthogonal projection upon the planes yz, zx, xy. The formula (1.2) is equally evident geometrically if S is a polyhedron. It is then clear that the formula (1.2) *should* result by approximating S by polyhedrons. At any rate, this is the point of view which is significant for the problem of PLATEAU. However, the situation is much more complicated than in the case of the length.

I.3. The situation can be strikingly illustrated by the famous example of H. A. SCHWARZ[2]. Let S be the surface

$$S: x^2 + y^2 = 1, \qquad 0 \leq z \leq 1.$$

---

[1] By a *general* curve we mean here one which is *not a polygon*. For an *actually* general continuous curve (1.1) is generally wrong. Cf. I.11.

[2] Gesammelte Mathematische Abhandlungen, vol. I pp. 309—311. We have slightly changed the notations of SCHWARZ.

Cut S along the generator $x = 1$, $y = 0$, $0 \leq z \leq 1$, and then spread S upon a plane. The result is a rectangle R with sides 1 and 2π. Hence $\mathfrak{A}(S) = 2\pi$. Subdivide the sides of R into m and n parts respectively. Subdivide R, by parallels to the sides through the points of division, into mn congruent rectangles r. Subdivide every one of these rectangles into two triangles by drawing a diagonal. Thus R is subdivided into a network of $2mn$ triangles. Bend R so as to obtain S, and use the vertices of the network as the vertices of an inscribed polyhedron. The area $\mathfrak{A}_{m,n}$ of this polyhedron is given by

$$\mathfrak{A}_{m,n} = 2n \sin \frac{\pi}{n},$$

and hence $\mathfrak{A}_{m,n} \to 2\pi = \mathfrak{A}(S)$ for $m, n \to \infty$, which is all right. Subdivide, however, every one of the rectangles r into four triangles by drawing both diagonals. There results an inscribed polyhedron, the area $\mathfrak{A}_{m,n}^{*}$ of which is given by

$$\mathfrak{A}_{m,n}^{*} = 2n \sin \frac{\pi}{2n} + \left[\frac{1}{4} + \frac{4 m^2}{n^4} \left(n \sin \frac{\pi}{2n} \right)^4 \right]^{\frac{1}{2}} \times 2n \sin \frac{\pi}{n}.$$

Since we used this time a finer subdivision, it might be expected that we get a better approximation, which is however obviously not the case. Indeed, if $m = n^3$, then $\mathfrak{A}_{m,n}^{*} \to \infty$. If $m = n$, then $\mathfrak{A}_{m,n}^{*} \to 2\pi = \mathfrak{A}(S)$. Since always

$$\mathfrak{A}_{m,n}^{*} \geq 2n \sin \frac{\pi}{2n} + n \sin \frac{\pi}{n},$$

it is clear that if $m, n \to \infty$ in any manner, then $\mathfrak{A}_{m,n}^{*}$ never converges to a limit $< 2\pi$. On the other hand it is obvious that every number k such that $2\pi \leq k \leq +\infty$ can be obtained as the limit of $\mathfrak{A}_{m,n}^{*}$, if m, n both go to infinity in a proper way.

Hence the area of inscribed polyhedrons, approximating a given surface S, do not converge, in general, to $\mathfrak{A}(S)$. This fact invalidates the geometrical interpretation of the formula (1.2) which was generally accepted before the example of SCHWARZ became known. A great number of new interpretations of (1.2) have since been proposed. In most cases, the idea of approximating the given surface by polyhedrons has been altogether dropped. However, as far as the problem of PLATEAU is concerned, the most essential facts concerning the area have been brought to light in efforts to clear up the relation between the area of a surface and the areas of approximating polyhedrons. We are going to give a brief account of the theory of the area from this point of view.

I.4. The first thing is to define the area of a surface. Of the many definitions which have been proposed only the definition given by LEBESGUE in his Thesis[1] became significant for the

---

[1] Intégrale, longueur, aire. Ann. Mat. pura appl. Vol. 7 (1902) pp. 231—359.

problem of PLATEAU, and therefore only that definition will be considered here[1].

In the example of SCHWARZ (see I.3) the areas of the approximating polyhedrons showed the tendency of converging to values larger than the area of the given surface S. The definition of the area given by LEBESGUE is based on the intuitive assumption that this tendency is absolutely general: if a sequence of surfaces converges to a surface, then the areas never converge to a value less than the area of the limit surface. Given then a class of surfaces S, we wish to define the area $\mathfrak{A}(S)$ of S in such a way that the above intuitive assumption be satisfied, that is to say in such a way that

$$\underline{\lim}\mathfrak{A}(S_n) \geqq \mathfrak{A}(S) \quad \text{if} \quad S_n \to S.$$

In other words, $\mathfrak{A}(S)$ has to be a lower semi-continuous functional. We also require that it must be possible to compute $\mathfrak{A}(S)$ by approximating S by polyhedrons; in other words, we require that there exists, for every surface S, a sequence of polyhedrons $\mathfrak{P}_n$ such that

$$\mathfrak{P}_n \to S \quad \text{and} \quad \mathfrak{A}(\mathfrak{P}_n) \to \mathfrak{A}(S).$$

Finally, we require that if S is a polyhedron, then $\mathfrak{A}(S)$ is equal to the area of the polyhedron in the elementary sense. These three conditions determine $\mathfrak{A}(S)$ univocally. Indeed, for every sequence of polyhedrons $\mathfrak{P}_n$ converging to S we must have $\underline{\lim}\mathfrak{A}(\mathfrak{P}_n) \geqq \mathfrak{A}(S)$, while the sign of equality holds for at least one sequence $\mathfrak{P}_n$. That is to say, $\mathfrak{A}(S)$ is the smallest value which is the limit of the areas of polyhedrons converging to S. This is the definition of the area given by LEBESGUE.

This definition, if it is to be consistent, implies the theorem that if a sequence of polyhedrons $\mathfrak{P}_n$ converges to a polyhedron $\mathfrak{P}$, then $\underline{\lim}\mathfrak{A}(\mathfrak{P}_n) \geqq \mathfrak{A}(\mathfrak{P})$, where $\mathfrak{A}$ denotes the area in the elementary sense (that is to say the sum of the areas of the faces of the polyhedron). Besides, the notions used in the definition must first be clearly defined. These points will be considered later on. For the moment, we wish to call a few peculiar facts to the attention of the reader.

I.5. Suppose S consists of the points in and on a JORDAN curve C situated in a plane. As is well known, the two-dimensional measure of C might be positive, and therefore the question arises as to whether $\mathfrak{A}(S)$ is the interior or the exterior measure. Since C can be approximated by polygons from the inside, it follows readily that $\mathfrak{A}(S)$ is at most equal to the interior area, that is to say to the measure of the open domain bounded by C.

Now, one of the most natural assumptions concerning the area is this: if a surface is projected orthogonally upon a plane, then the area

---

[1] For literature and a systematic presentation, see T. RADÓ: Über das Flächenmaß rektifizierbarer Flächen. Math. Ann. Vol. 100 (1928) pp. 445—479.

of the surface is at least equal to the measure of the projection. In the above example, the projection is the closed region bounded by C. Hence, if the two-dimensional measure of C is positive, we have an example showing that *the area of a surface is in general less than the measure of the orthogonal projection of the surface upon a plane.*

I.6. This situation, which is an inevitable consequence of the requirement that $\mathfrak{A}(S)$ be a lower semi-continuous functional, constitutes one of the main difficulties in handling the definition of LEBESGUE. GEÖCZE devised the following simple example which shows the situation possibly at its worst[1]. Let the surface S be given by equations

$$S: x = x(u,v), \quad y = y(u,v), \quad z = z(u,v), \quad 0 \leq u \leq 1, \quad 0 \leq v \leq 1,$$

where $x(u,v), y(u,v), z(u,v)$ are continuous. Subdivide the square $0 \leq u \leq 1, 0 \leq v \leq 1$ into n^2 congruent squares, and subdivide every one of these smaller squares into two triangles by drawing a diagonal. Use the points of S which correspond to the vertices of this triangular net as the vertices of an inscribed polyhedron $\mathfrak{P}_n$. Then, by definition,

$$\mathfrak{A}(S) \leq \varliminf \mathfrak{A}(\mathfrak{P}_n) \quad \text{for} \quad n \to \infty.$$

Suppose now that $x(u,v), y(u,v), z(u,v)$ are functions of u alone:

$$x(u,v) \equiv f_1(u), \quad y(u,v) \equiv f_2(u), \quad z(u,v) \equiv f_3(u).$$

Obviously, $\mathfrak{A}(\mathfrak{P}_n) = 0$ for every n, and hence $\mathfrak{A}(S) = 0$, which looks all right, since S reduces in reality to the curve

$$\Gamma: x = f_1(u), \quad y = f_2(u), \quad z = f_3(u).$$

If we choose however Γ as a PEANO curve filling a cube, then we obtain an example showing that *a surface might contain every point of a cube and might still have a zero area.*

I.7. The definition of LEBESGUE implies a previous definition of convergent sequences of surfaces. This latter definition will be based on the notion of the distance of two surfaces. We shall now show in a simple example how important the definition of the distance is. Take two surfaces S_1, S_2. Define the distance of S_1 and S_2 as the smallest number δ with the properties: 1. for every point P_1 of S_1 there exists a point P_2 of S_2 such that the distance $P_1 P_2$ is less than or equal to δ, and 2. for every point P_2 of S_2 there exists a point P_1 of S_1 such that the distance $P_1 P_2$ is less than or equal to δ. Given then a sequence of surfaces S_n and a surface S, $S_n \to S$ means that the distance of S_1 and S_2 converges to zero.

Suppose we use this definition of convergence in the definition of the area (which we shall not do). Given then a continuous surface S, and an $\varepsilon > 0$, it is clear that we can take a very long and very narrow

---

[1] Z. DE GEÖCZE: Sur l'exemple d'une surface dont l'aire est égale à zéro et qui remplit un cube. Bull. Soc. Math. France (1913) pp. 29—31.

ribbon of paper, of rectangular shape and with an area less than ε, and deform the ribbon so as to obtain a polyhedron the distance of which from S, in the sense defined above, is less than ε. Doing this for $\varepsilon = 1, \frac{1}{2}, \frac{1}{3}, \ldots$, we obtain a sequence of polyhedrons $\mathfrak{P}_n$ such that $\mathfrak{A}(\mathfrak{P}_n) < \frac{1}{n}$ and $\mathfrak{P}_n \to S$. Consequently, $\mathfrak{A}(S) = 0$. That is to say, if we use the definition of the distance given above (which we shall not do), then the area of every continuous surface is zero.

This shows clearly that *the definition of convergent sequences of surfaces is of the greatest importance for the definition of the area* given by LEBESGUE.

I.8. We now shall give the exact definition of $\mathfrak{A}(S)$ for the class of the *continuous surfaces of the type of the circular disc*. Such a surface is defined by a set of equations

$$S : x = x(u, v), \qquad y = y(u, v), \qquad z = z(u, v), \qquad (u, v) \text{ in } R,$$

where R denotes some JORDAN region (that is to say the set of points in and on a JORDAN curve), and $x(u, v), y(u, v), z(u, v)$ are continuous in R. We do not suppose that distinct points (u, v) are carried into distinct points (x, y, z). Given then two such surfaces

$$S_1 : x = x_1(u, v), \qquad y = y_1(u, v), \qquad z = z_1(u, v), \qquad (u, v) \text{ in } R_1, \qquad (1.3)$$

and

$$S_2 : x = x_2(u, v), \qquad y = y_2(u, v), \qquad z = z_2(u, v), \qquad (u, v) \text{ in } R_2, \qquad (1.4)$$

consider a topological correspondence T between R_1 and R_2. Denote by $(u_1, v_1), (u_2, v_2)$ a couple of corresponding points, and denote by P_1, P_2 the points which correspond to $(u_1, v_1), (u_2, v_2)$ by means of the equations (1.3), (1.4) respectively. The maximum of the distance $P_1 P_2$, for all possible positions of the points $(u_1, v_1), (u_2, v_2)$ corresponding to each other under T, will be denoted by $\delta(T)$. The greatest lower bound of $\delta(T)$, for all possible topological correspondences between R_1 and R_2, is *the distance, in the sense of* FRÉCHET, *of* S_1 *and* S_2 and will be denoted by $d(S_1, S_2)$[1].

If $d(S_1, S_2) = 0$, the surfaces S_1, S_2 will be considered as identical. We shall say also, if $d(S_1, S_2) = 0$, that (1.3) and (1.4) are *parametric representations of the same surface*.

If we have a surface S and a sequence of surfaces S_n, such that $d(S_n, S) \to 0$, then we shall say that $S_n \to S$.

It should be observed that the definition of the distance, in the sense of FRÉCHET, presupposes that we are dealing with surfaces of the same topological type. In this report, except for parts of Chapter VI, we shall consider continuous surfaces of the type of the circular disc only. The term *surface* will be used in this sense, unless the contrary is explicitly stated.

---

[1] Sur la distance de deux surfaces. Ann. Soc. Polon. math. Vol. 3 (1924) pp. 4—19.

It is also important to remark that two surfaces S_1, S_2 might consist of the same points (x, y, z) without being identical in the sense $d(S_1, S_2) = 0$.

I.9. A surface S will be called a *polyhedron* and will be denoted by $\mathfrak{P}$ if it admits of a parametric representation

$$\mathfrak{P} : x = x(u, v), \qquad y = y(u, v), \qquad z = z(u, v), \qquad (u, v) \text{ in } R, \qquad (1.5)$$

with the following properties. The JORDAN region R can be subdivided into a finite number of non-overlapping curvilinear triangles $\delta_1, \delta_2, \ldots, \delta_n$ in such a way that every one of these triangles is carried, by the equations (1.5), in a one-to-one and continuous way into a (non-degenerate) plane rectilinear triangle in the xyz-space. The boundary of R is carried in a one-to-one and continuous way into a simple closed polygon. Such a representation will be called a *typical representation* of $\mathfrak{P}$.

If $\varDelta_1, \varDelta_2, \ldots, \varDelta_n$ are the (plane and rectilinear) triangles into which $\delta_1, \delta_2, \ldots, \delta_n$ are carried by the equations (1.5), then the sum of the areas of $\varDelta_1, \varDelta_2, \ldots, \varDelta_n$ will be called the elementary area of $\mathfrak{P}$ and will be denoted by $E(\mathfrak{P})$. Then $E(\mathfrak{P})$ can be shown to be independent of the choice of the typical representation which has been used for the computation.

I.10. Given a surface S, consider a sequence $\mathfrak{P}_n$ of polyhedrons converging, in the FRÉCHET sense (see I.8) to S. Consider $\underline{\lim} E(\mathfrak{P}_n)$. Then the greatest lower bound of $\underline{\lim} E(\mathfrak{P}_n)$, for all possible sequences $\mathfrak{P}_n \to S$, is by definition the area (in the sense of LEBESGUE) of S, and will be denoted by $\mathfrak{A}(S)$. Thus $\mathfrak{A}(S)$ is defined for every continuous surface of the type of the circular disc. $\mathfrak{A}(S)$ might be $+\infty$.

This definition of $\mathfrak{A}(S)$ is consistent as it stands, and would be logically consistent even if it would not be true that if S is a polyhedron $\mathfrak{P}$, then $\mathfrak{A}(\mathfrak{P}) = E(\mathfrak{P})$. On the other hand it is clear that $\mathfrak{A}(\mathfrak{P}) = E(\mathfrak{P})$ must be true if the definition is to serve any useful purpose. The theorem $\mathfrak{A}(\mathfrak{P}) = E(\mathfrak{P})$ is true[1], but not obvious; its truth, and in a general way the usefulness of $\mathfrak{A}(S)$, depends essentially upon the definition of the distance of two surfaces, which we have decided to use (see I.7).

I.11. In developing the theory of the area $\mathfrak{A}(S)$, a good working program is obtained by setting up the principle that the theory of the area should be analogous to the theory of the length. Let us recall therefore a few facts concerning the length. If

$$C : x = x(t), \qquad y = y(t), \qquad z = z(t), \qquad a \leq t \leq b$$

are the equations of a continuous curve, then the length $l(C)$ is defined as the least upper bound of the lengths of inscribed polygons. While this definition is not analogous to the definition of $\mathfrak{A}(S)$, it can easily

---

[1] For a direct proof, see M. FRÉCHET: Sur la semi-continuité en géométrie élémentaire, Nouvelles Ann. de Math. Vol. 3 (1924).

be shown that it is equivalent to the following definition. Define first the distance, in the FRÉCHET sense, of two continuous curves

$$C_1 : x = x_1(t), \qquad y = y_1(t), \qquad z = z_1(t), \qquad a_1 \leq t \leq b_1,$$
$$C_2 : x = x_2(t), \qquad y = y_2(t), \qquad z = z_2(t), \qquad a_2 \leq t \leq b_2$$

by using one-to-one and continuous correspondences between the intervals $a_1 \leq t \leq b_1$ and $a_2 \leq t \leq b_2$ in exactly the same manner as the distance of two continuous surfaces has been defined in I.8. Then the length $l(C)$ of a continuous curve C can be defined as the smallest number which is the limit of the lengths of polygons converging, in the FRÉCHET sense, to C.

On account of this definition, $l(C)$ clearly is a lower semi-continuous functional. That is to say, if $C_n \to C$, then $\underline{\lim} l(C_n) \geq l(C)$. For a polygon $\mathfrak{p}$, $l(\mathfrak{p})$ is equal to the length of $\mathfrak{p}$ in the elementary sense. If a sequence $\mathfrak{p}_n$ of polygons converges to a continuous curve C, then $\underline{\lim} l(\mathfrak{p}_n) \geq l(C)$. If $\mathfrak{p}_n \to C$, and $\mathfrak{p}_n$ is inscribed in C, then $\lim l(\mathfrak{p}_n) = l(C)$. This gives a convergent process for calculating $l(C)$. Let

$$C : x = x(t), \qquad y = y(t), \qquad z = z(t), \, a \leq t \leq b$$

be the equations of C. Subdivide the interval $a \leq t \leq b$ by points

$$a = t_0 < t_1 < \cdots < t_{k-1} < t_k < \cdots < t_n = b$$

into subintervals. Denote by D this subdivision and by $\|D\|$ the greatest length of any of the subintervals. Then

$$\lim \sum_{k=1}^{n} \{[x(t_k) - x(t_{k-1})]^2 + [y(t_k) - y(t_{k-1})]^2 + [z(t_k) - z(t_{k-1})]^2\}^{\frac{1}{2}} = l(C) \quad (1.6)$$

if $\|D\| \to 0$.

The relation between $l(C)$ and the classical integral formula is also completely known[1]. If $l(C)$ is finite, then the coordinate functions $x(t), y(t), z(t)$ are of bounded variation. The differential coefficients $x'(t), y'(t), z'(t)$ exist almost everywhere, and

$$\int_a^b (x'^2 + y'^2 + z'^2)^{\frac{1}{2}} dt \leq l(C).$$

The sign of equality holds if and only if $x(t), y(t), z(t)$ are absolutely continuous. There always exist representations of C which satisfy this condition [provided $l(C)$ is finite]. For instance, if $s(t)$ denotes the length of the arc corresponding to the interval from a to t, then x, y, z as functions of $s = s(t)$ satisfy even the LIPSCHITZ condition.

I.12. With the preceding facts as a basis of comparison, the theory of the area $\mathfrak{A}(S)$ appears as being surprisingly incomplete. $\mathfrak{A}(S)$ easily is found to be a lower semi-continuous functional, and we also mentioned

---

[1] Cf. L. TONELLI: Sopra alcuni polinomi approssimativi. Ann. Mat. pura appl. Vol. 25 (1916).

that for polyhedrons $\mathfrak{P}$ the area $\mathfrak{A}(\mathfrak{P})$ is equal to the area in the elementary sense. If a sequence of polyhedrons $\mathfrak{P}_n$ converges to a surface S, then $\varliminf \mathfrak{A}(\mathfrak{P}_n) \geqq \mathfrak{A}(S)$. But no general procedure is known which would permit us to construct, for every continuous surface S of the type of the circular disc, a sequence of polyhedrons $\mathfrak{P}_n$ such that $\mathfrak{P}_n \to S$ and $\mathfrak{A}(\mathfrak{P}_n) \to \mathfrak{A}(S)$, although such a sequence certainly exists by definition. It is not known if it always is possible to select a sequence $\mathfrak{P}_n$ of *inscribed* polyhedrons such that $\mathfrak{P}_n \to S$ and $\mathfrak{A}(\mathfrak{P}_n) \to \mathfrak{A}(S)$. No general convergent process is known which would permit us to compute $\mathfrak{A}(S)$ in terms of the coordinate functions $x(u, v)$, $y(u, v)$, $z(u, v)$, while formula (1.6) in I.11 gives such a process for the computation of the length of every continuous curve. It is not known if for every surface S with a finite $\mathfrak{A}(S)$ there exists a representation such that the integral (1.2) exists and is equal to $\mathfrak{A}(S)$.

On the other hand, very satisfactory results have been obtained for certain special classes of surfaces which now will be considered briefly.

I.13. Suppose S is given by an equation

$$S : z = z(x, y), \qquad (x, y) \text{ in } R,$$

where $z(x, y)$ is single-valued and continuous in the closed region R which we shall suppose to be a rectangle

$$R : a \leqq x \leqq b, \quad c \leqq y \leqq d.$$

The following expressions, introduced by the Hungarian mathematician GEÖCZE[1], are fundamental in the theory of $\mathfrak{A}(S)$. Let

$$r : x' \leqq x \leqq x'', \quad y' \leqq y \leqq y''$$

denote a rectangle r comprised in R. Put:

$$\alpha(z, r) = \int_{x'}^{x''} |z(\xi, y'') - z(\xi, y')| \, d\xi,$$

$$\beta(z, r) = \int_{y'}^{y''} |z(x'', \eta) - z(x', \eta)| \, d\eta,$$

$$\gamma(r) = (x'' - x')(y'' - y'),$$

$$g(z, r) = (\alpha^2 + \beta^2 + \gamma^2)^{\frac{1}{2}}.$$

Consider then a subdivision D of R into rectangles, obtained by drawing a finite number of parallels to the x and y axes respectively. Denote by $\|D\|$ the greatest diagonal of the rectangles of D. Put

$$G(z, D) = \sum_{(r)} g(z, r),$$

---

[1] Quadrature des surfaces courbes. Math. naturwiss. Ber., Ungarn Vol. 26 (1910) pp. 1—88.

where the summation is extended over all the rectangles r of D. Then $G(z, D)$ will be called a GEÖCZE sum for the surface S.

GEÖCZE proved the following facts concerning these sums. If D^* is a subdivision of D, then $G(z, D) \leq G(z, D^*)$. If $z_n(x, y) \rightrightarrows z(x, y)$ in R, then $G(z_n, D) \rightarrow G(z, D)$ for every fixed subdivision D. For every subdivision D we have the inequality $G(z, D) \leq \mathfrak{A}(S)$. If D_n is a sequence of subdivisions such that $\|D_n\| \rightarrow 0$, then $G(z, D_n)$ converges to a definite limit independent of the particular choice of the sequence D_n. Put

$$\Gamma(z) = \lim G(z, D) \quad \text{for} \quad \|D\| \rightarrow 0.$$

$\mathfrak{A}(S)$ is finite if and only if $\Gamma(z)$ is finite.

GEÖCZE also stated the theorem $\Gamma(z) = \mathfrak{A}(z)$. He proved it however only in the special case when $z(x, y)$ satisfies the LIPSCHITZ condition. In this case he proved also that

$$G(z, D) \rightarrow \iint_R (1 + p^2 + q^2)^{\frac{1}{2}} dx\, dy \quad \text{for} \quad \|D\| \rightarrow 0,$$

where the partial derivatives $p = z_x$, $q = z_y$ exist almost everywhere and are bounded and measurable, on account of the LIPSCHITZ condition. Thus GEÖCZE proved that if $z(x, y)$ satisfies the LIPSCHITZ condition, then $\mathfrak{A}(S)$ is given by the classical formula

$$\mathfrak{A}(S) = \iint_R (1 + p^2 + q^2)^{\frac{1}{2}} dx\, dy.$$

Suppose now only that $\mathfrak{A}(S)$ is finite. Then TONELLI[1] proved that the partial derivatives $p = z_x$ and $q = z_y$ exist almost everywhere in R, that $(1 + p^2 + q^2)^{\frac{1}{2}}$ is summable, and that

$$\iint_R (1 + p^2 + q^2)^{\frac{1}{2}} dx\, dy \leq \mathfrak{A}(S).$$

TONELLI obtained then the beautiful theorem that the sign of equality holds if and only if $z(x, y)$ is absolutely continuous[2]. By a combination of the methods of GEÖCZE and TONELLI, RADÓ[3] proved the theorem, stated without proof by GEÖCZE, that

$$G(z, D) \rightarrow \mathfrak{A}(S) \quad \text{for} \quad \|D\| \rightarrow 0, \tag{1.6.1}$$

under the assumption only that $z(x, y)$ is continuous. The theorem holds even if $\mathfrak{A}(S)$ is infinite. (1.6.1) gives a convergent process for

[1] Sulla quadratura delle superficie. Rend. Accad. naz. Lincei Vol. 3 (1926) pp. 445—450.

[2] L. TONELLI: Sulla quadratura delle superficie. Rend. Accad. naz. Lincei Vol. 3 (1926) pp. 633—638.

[3] Sur le calcul de l'aire des surfaces courbes. Fundam. Math. Vol. 10 (1927) pp. 197—210. A similar result has been obtained later by TONELLI: Sulla quadratura delle superficie. Rend. Accad. naz. Lincei Vol. 5 (1927) pp. 313—318.

the area $\mathfrak{A}(S)$ of every continuous surface S of the form $z = z(x, y)$, and this process, as far as the appearance of the formulas is concerned, is absolutely analogous to the process given by (1.6) in I.11 for the length of a curve. It should be observed however that $G(z, D)$ cannot be interpreted as the area of a polyhedron. The problem of constructing a sequence $\mathfrak{P}_n$ of polyhedrons such that $\mathfrak{P}_n \to S$ and $\mathfrak{A}(\mathfrak{P}_n) \to \mathfrak{A}(S)$ is still unsolved even if stated only for continuous surfaces of the special form $z = z(x, y)$.

Since (1.6.1) gives an analytic expression for $\mathfrak{A}(S)$, it might be expected that the theory of $\mathfrak{A}(S)$ can be based solely on that analytic expression and is thus accessible to general analytic methods, without further reference to the geometrical definition of $\mathfrak{A}(S)$. This program has been stated and carried out by S. SAKS[1]. The basic theorem in his work is as follows. Let (x_0, y_0) be an interior point of the rectangle R, and denote by σ a small square comprising (x_0, y_0). Denote by S_σ the portion of S situated above σ. Then, if the diameter of σ converges to zero,

$$\frac{\mathfrak{A}(S_\sigma)}{\text{area of } \sigma} \to [1 + p(x_0, y_0)^2 + q(x_0, y_0)^2]^{\frac{1}{2}}$$

for almost every point (x_0, y_0) in R (that is to say for every point in R with the possible exception of a set of measure zero). It is presupposed in the theorem that $\mathfrak{A}(S)$ is finite.

I.14. For surfaces of the general type

$$S: x = x(u, v), \quad y = y(u, v), \quad z = z(u, v), \quad (u, v) \text{ in } R, \quad (1.6.2)$$

the theory of $\mathfrak{A}(S)$ is very incomplete. The first significant result is again due to GEÖCZE, and is concerned with the so-called rectifiable surfaces. This very misleading term is used to denote surfaces S which admit of a representation (1.6.2) such that $x(u, v)$, $y(u, v)$, $z(u, v)$ satisfy the LIPSCHITZ condition in R. Such a representation will be called a *typical representation of the rectifiable surface S*. GEÖCZE proved[2] that for every rectifiable surface S, given in typical representation, we have

$$\mathfrak{A}(S) = \iint\limits_R \left[\left(\frac{\partial(y, z)}{\partial(u, v)} \right)^2 + \left(\frac{\partial(z, x)}{\partial(u, v)} \right)^2 + \left(\frac{\partial(x, y)}{\partial(u, v)} \right)^2 \right]^{\frac{1}{2}} du\, dv.$$

The rectifiable surfaces have also been studied by RADEMACHER[3]. Suppose, for the sake of simplicity, that R is the square

$$R: 0 \leqq u \leqq 1, \quad 0 \leqq v \leqq 1.$$

---

[1] Sur l'aire des surfaces $z = f(x, y)$. Acta Litt. Sci. Szeged Vol. 3 (1927) pp. 170—176.

[2] This part of the work of GEÖCZE has only been published in Hungarian. For references, and for a simplified presentation of the results of GEÖCZE, see T. RADÓ: Über das Flächenmaß rektifizierbarer Flächen. Math. Ann. Vol. 100 (1928) pp. 445—479.

[3] Über partielle und totale Differenzierbarkeit I, II. Math. Ann. Vol. 79 (1919) pp. 340—359 and Vol. 81 (1920) pp. 52—63.

Subdivide R into a finite number of non-overlapping (rectilinear) triangles. Denote by $\mathfrak{N}$ this triangular net, by $\|\mathfrak{N}\|$ the greatest side of the triangles of $\mathfrak{N}$, by φ the smallest angle occurring in the triangles of $\mathfrak{N}$, and by $\mathfrak{P}$ the polyhedron whose vertices are the points of the surface S corresponding to the vertices of $\mathfrak{N}$ by means of the equations (1.6.2) of the rectifiable surface S given in typical representation. Then RADEMACHER proves that $\mathfrak{A}(\mathfrak{P})$ converges to the classical integral if $\|\mathfrak{N}\| \to 0$, provided φ remains larger than a fixed positive number. A more general criterion is obtained in terms of the greatest angle $\varPhi$ occurring in the triangles of $\mathfrak{N}$, namely the criterion that $\mathfrak{A}(\mathfrak{P})$ converges to the classical integral for $\|\mathfrak{N}\| \to 0$, provided $\pi - \varPhi$ remains larger than a fixed positive number. This criterion has been stated, for the elementary case when $x(u, v)$, $y(u, v)$, $z(u, v)$ have continuous derivatives, by FRÉCHET[1]. The methods of RADEMACHER permit us to generalize the criterion to rectifiable surfaces given in typical representation. Thus the problem of determining a sequence of polyhedrons $\mathfrak{P}_n$ such that $\mathfrak{P}_n \to S$ and $\mathfrak{A}(\mathfrak{P}_n) \to \mathfrak{A}(S)$ is solved for rectifiable surfaces S. The method of RADEMACHER is based on the fact that a rectifiable surface S has a definite tangent plane almost everywhere and that the restrictions upon the net $\mathfrak{N}$ imply that the faces of the corresponding inscribed polyhedron $\mathfrak{P}$ approximate the tangent planes of S.

I.15. This situation and the surmised analogy with the theory of the length might suggest a generalization whose impossibility however is demonstrated by a curious remark of S. SAKS to the effect that *a surface S with finite area $\mathfrak{A}(S)$ need not have a tangent plane anywhere*[2], not even if the surface admits a representation of the very special type $z = z(x, y)$. The reason for this might be explained as follows. Let the surface S coincide with the square $0 \leqq x \leqq 1, 0 \leqq y \leqq 1$, $z \equiv 0$. Take an interior point (x_0, y_0), a small $\varrho > 0$, and replace the circular disc $(x - x_0)^2 + (y - y_0)^2 \leqq \varrho^2$ by a cone of revolution of height h. No matter how large h is, the area of this cone will be arbitrary small if ϱ is sufficiently small. Thus we obtain from S a new surface the area of which is very slightly larger than the area of S, while the smoothness of S has been badly disturbed around (x_0, y_0). The example of S. SAKS is then obtained by putting on sharper and sharper cones and still keeping the area finite. The process is similar to that which gives continuous curves without tangents[3], with the essential difference that in the case of a curve a disturbance of the smoothness appreciably

[1] Note on the area of a surface. Proc. London Math. Soc. Vol. 24 (1926).

[2] The following is based on an oral communication of S. SAKS. — (*Added in proof.*) For the details, see his paper, On the surfaces without tangent planes, Ann. of Math. Vol. 34 (1933) pp. 114—124.

[3] See for instance K. KNOPP: Ein einfaches Verfahren etc. Math. Z. Vol. 2 (1918) pp. 1—26.

increases the length and therefore the total disturbance of the smoothness drives the length to infinity.

The long thin cones with small areas are responsible for another serious evil which will be discussed in III.13.

I.16. Important contributions to the theory of the area are due to recent investigations of E. J. McShane[1] and of C. B. Morrey[2]. We restrict ourselves to state just one theorem which will be needed later. Let there be given a continuous surface

$$S : x = x(u, v), \quad y = y(u, v), \quad z = z(u, v), \quad u^2 + v^2 \leqq 1.$$

Suppose that

1. $x(u, v)$, $y(u, v)$, $z(u, v)$ are absolutely continuous functions of u for almost every v and absolutely continuous functions of v for almost every u and

2. the Dirichlet integrals

$$\iint (x_u^2 + x_v^2), \quad \iint (y_u^2 + y_v^2), \quad \iint (z_u^2 + z_v^2),$$

taken over $u^2 + v^2 < 1$, are finite.

Then the area of S is given by the classical integral formula (1.2).

I.17. We shall also need, in Chapter VI, a result of McShane[3] concerning the so-called inequality of Steiner. Denote by R a Jordan region in the xy-plane, and consider two continuous surfaces

$$S_1 : z = z_1(x, y), \quad S_2 : z = z_2(x, y), \quad (x, y) \text{ in } R.$$

Denote by S the surface

$$S : z = \tfrac{1}{2}(z_1 + z_2), \quad (x, y) \text{ in } R.$$

Then

$$\mathfrak{A}(S) \leqq \tfrac{1}{2}(\mathfrak{A}(S_1) + \mathfrak{A}(S_2)). \tag{1.7}$$

This is the inequality of Steiner. It is obvious for polyhedrons, and a passage to the limit proves it for the general case. It is important for the sequel to discuss the sign of equality. Let us restrict ourselves to the case when R is a rectangle with sides parallel to the axes. Subdivide R into smaller rectangles r of the same type. For every r we have then the inequality

$$\mathfrak{A}(S^{(r)}) \leqq \tfrac{1}{2}(\mathfrak{A}(S_1^{(r)}) + \mathfrak{A}(S_2^{(r)})), \tag{1.8}$$

where $S^{(r)}$, $S_1^{(r)}$, $S_2^{(r)}$ denote the portions of S, S_1, S_2 corresponding to r. Adding up these inequalities, it follows that the sign of equality in

---

[1] Integrals over surfaces in parametric form. Bull. Amer. Math. Soc. Vol. 38 (1932) p. 810.

[2] A class of representations of manifolds. Bull. Amer. Math. Soc. Vol. 38 (1932) p. 809. An analytic criterion that a surface possess finite Lebesgue area. Bull. Amer. Math. Soc. Vol. 39 (1933) p. 41.

[3] On a certain inequality of Steiner: Ann. of Math. Vol. 33 (1932) pp. 123—138.

(1.7) holds if and only if it holds in (1.8) for every rectangle r comprised in R, or, in other words, if and only if the function of rectangles

$$F(r) = \tfrac{1}{2}\left(\mathfrak{A}(S_1^{(r)}) + \mathfrak{A}(S_2^{(r)})\right) - \mathfrak{A}(S^{(r)})$$

vanishes identically. Take then any point (x, y) interior to R and let r be any small square, with sides parallel to the axes, comprising (x, y). Denote by $a(r)$ the area of r. Then, on account of the theorem of S. Saks, stated at the end of I.13,

$$\frac{F(r)}{a(r)} \to \frac{1}{2}\left((1 + p_1^2 + q_1^2)^{\frac{1}{2}} + (1 + p_2^2 + q_2^2)^{\frac{1}{2}}\right) - (1 + p^2 + q^2)^{\frac{1}{2}} \qquad (1.9)$$

for almost every (x, y) in R. As $F(r) \equiv 0$, it follows that the right-hand side of (1.9) vanishes almost everywhere in R. As

$$p = \tfrac{1}{2}(p_1 + p_2), \quad q = \tfrac{1}{2}(q_1 + q_2),$$

this implies that $p_1 = p_2$, $q_1 = q_2$ almost everywhere in R. That is to say: *if the sign of equality holds in* (1.7), *then* $p_1 - p_2 = 0$, $q_1 - q_2 = 0$ *almost everywhere in* R.

From this it does *not* follow, in general, that $z_2 - z_1$ is constant in R. An important case in which this conclusion is legitimate is the case when $\mathfrak{A}(S_1)$ and $\mathfrak{A}(S_2)$ are both given by the classical integral formulas

$$\mathfrak{A}(S_1) = \iint (1 + p_1^2 + q_1^2)^{\frac{1}{2}},$$

$$\mathfrak{A}(S_2) = \iint (1 + p_2^2 + q_2^2)^{\frac{1}{2}}.$$

Indeed, according to Tonelli (see I.13), z_2 and z_1 are then both absolutely continuous, and so is therefore $z_2 - z_1$. The vanishing of $p_2 - p_1$, $q_2 - q_1$ almost everywhere in R actually implies then that $z_2 - z_1$ is constant.

I.18. In the theory of minimal surfaces, conformal mapping plays an important part. We are going to recall a few significant facts concerned with such maps. Consider a surface

$$S : x = x(u, v), \quad y = y(u, v), \quad z = z(u, v), \quad u^2 + v^2 \le 1. \qquad (1.10)$$

Take an interior point (u_0, v_0), and map a vicinity V_0 of (u_0, v_0) upon a vicinity V_0^* of some point (α_0, β_0) in the $\alpha\beta$-plane in a one-to-one and continuous way by equations $\alpha = \alpha(u, v)$, $\beta = \beta(u, v)$. Suppose that x, y, z, as functions of α, β, have continuous partial derivatives of the first order in V_0^* which satisfy there the relations $E^* = G^*$, $F^* = 0$, where

$$E^* = x_\alpha^2 + y_\alpha^2 + z_\alpha^2,$$
$$F^* = x_\alpha x_\beta + y_\alpha y_\beta + z_\alpha z_\beta,$$
$$G^* = x_\beta^2 + y_\beta^2 + z_\beta^2.$$

Then we shall say that α, β are *local isothermic parameters*. If $E^* = G^* > 0$ in V_0^*, then the correspondence between V_0^* and S is one-to-one and conformal, that is to say angles are preserved under the correspondence.

In the sequel, we shall have to permit the vanishing of $E^* = G^*$, that is to say we shall use the term isothermic in a more general sense than is usual in differential geometry.

The most general condition, known at present, for the existence of local isothermic parameters is as follows[1]. The surface S, given by (1.10), admits of local isothermic parameters if the following conditions are satisfied.

1. $x(u, v)$, $y(u, v)$, $z(u, v)$ have continuous partial derivatives of the first order.

2. The partial derivatives of the first order satisfy, in every closed region R entirely interior to the unit circle, the LIPSCHITZ-HÖLDER condition with some exponent λ[*], possibly depending upon R, such that $0 < \lambda < 1$.

3. $EG - F^2 > 0$ for $u^2 + v^2 < 1$, where

$$E = x_u^2 + y_u^2 + z_u^2,$$
$$F = x_u x_v + y_u y_v + z_u z_v,$$
$$G = x_v^2 + y_v^2 + z_v^2.$$

For many applications it would be important to replace 1. and 2. by some less restrictive condition. The least restrictive condition which can be substituted for 1. and 2., without hurting the validity of the conclusion, is not yet known[2]. From the point of view of the problem of PLATEAU, as it appears at the time being, condition 3. is however the most disturbing, inasmuch as we shall have to consider surfaces which possibly do not satisfy $EG - F^2 > 0$. It is easy to see that for such surfaces local isothermic parameters will *not* exist in general, no matter how good the coordinate functions are. Consider for instance the case

$$S: x = u, \quad y = 0, \quad z = 0, \quad u^2 + v^2 \leq 1.$$

Suppose we have local isothermic parameters. Using the same notations as at the beginning of the present section I.18, we have then

$$\iint\limits_{V_0} (EG - F^2)^{\frac{1}{2}} = \iint\limits_{V_0^*} (E^* G^* - F^{*2})^{\frac{1}{2}},$$

---

[1] L. LICHTENSTEIN: Beweis des Satzes etc. Abh. preuß. Akad. Wiss. 1911. — A. KORN: Zwei Anwendungen der Methode der sukzessiven Annäherungen (Schwarz-Festschrift), p. 215—229. Berlin, Julius Springer 1914.

[*] A function $f(u, v)$ satisfies in R the LIPSCHITZ-HÖLDER condition with the exponent λ if $|f(u_1, v_1) - f(u_2, v_2)| \leq \gamma d^\lambda$, where d is the distance of the points (u_1, v_1), (u_2, v_2) and γ is a constant.

[2] The reviewer had the privilege to get information about yet unpublished investigations of J. E. McSHANE and of C. B. MORREY which lead to very general results. — In connection with his work on the problem of PLATEAU, T. RADÓ used maps which are conformal in a certain approximate sense and announced further developments on the subject [Bull. Amer. Math. Soc. Vol. 38 (1932) p. 129].

on account of the invariance of the area-integral. Since

$$E = 1, \quad F = 0, \quad G = 0, \quad E^* = G^*, \quad F^* = 0,$$

it follows that

$$\iint\limits_{V_0^*} E^* = \iint\limits_{V_0^*} G^* = 0,$$

and consequently $E^* \equiv 0$, $G^* \equiv 0$ in V_0^*. Hence $x_\alpha, y_\alpha, z_\alpha, x_\beta, y_\beta, z_\beta$ vanish identically in V_0^*, and thus x, y, z are constant in V_0^* and consequently in V_0, which however certainly is not the case.

I.19. As a substitute for the generally lacking conformal maps of surfaces which are not known to satisfy the condition $EG - F^2 > 0$, the fact that every polyhedron admits of a conformal map, in a certain generalized sense, has been used in the theory of the problem of PLATEAU[1]. We have namely the following theorem[2].

Every polyhedron $\mathfrak{P}$ admits of a typical representation (see I.9)

$$\mathfrak{P} : x = x(u, v), \quad y = y(u, v), \quad z = z(u, v), \quad u^2 + v^2 \leqq 1$$

with the following additional properties.

1. The sides of the (curvilinear) triangles $\delta_1, \delta_2, \ldots, \delta_n$ are analytic curves, including end-points.

2. None of the angles of these triangles is zero.

3. $x(u, v)$, $y(u, v)$, $z(u, v)$ are analytic in the interior of every one of the triangles $\delta_1, \delta_2, \ldots, \delta_n$ and satisfy there the equations $E = G$, $F = 0$.

Applications of this theorem will be considered in Chapter V and Chapter VI.

I.20. We shall review now certain facts, concerning curves, which will be used in the sequel. It is sufficient for our purposes to consider JORDAN arcs and JORDAN curves. A JORDAN arc is the one-to-one and continuous image of a closed interval $a \leqq t \leqq b$. A JORDAN curve is the one-to-one and continuous image of the unit circle $u^2 + v^2 = 1$.

Given two JORDAN arcs C_1 and C_2, their *distance* $d(C_1, C_2)$, in the sense of FRÉCHET, is defined as follows. Consider a topological correspondence T between C_1 and C_2, and denote by $M(T)$ the maximum distance of corresponding points. The greatest lower bound of $M(T)$, for all possible topological correspondences T, is by definition the distance $d(C_1, C_2)$. The definition of the distance of two JORDAN curves is exactly the same. Convergent sequences of JORDAN arcs and JORDAN curves are then defined in an obvious way in terms of the distance.

---

[1] See V.20.

[2] This theorem has been first proved by H. A. SCHWARZ: Über einen Grenzübergang durch alternierendes Verfahren. Gesammelte Mathematische Abhandlungen, Vol. II, pp.133—143. The theorem is also comprised as a special case in the fundamental theorem of uniformisation. See for instance CARATHÉODORY, Conformal representation (Cambridge Tracts 28), Chapter VII.

It should be observed that a JORDAN arc and a JORDAN curve have no distance in the sense of FRÉCHET, since there is no topological correspondence between them to start with.

I.21. A JORDAN arc C, by definition, admits of a representation

$$C: x = x(t), \quad y = y(t), \quad z = z(t), \quad a \leq t \leq b,$$

where these equations define a one-to-one and continuous correspondence between $a \leq t \leq b$ and C. It will be necessary to consider improper representations also, such that while t describes the interval, the corresponding point moves on C always in the same sense, but possibly with stops and jumps. Since such a representation obviously does not determine C, we shall rather speak of monotonic transformations, all the more because the term points to the source of the significant properties we shall want. The following definitions will be used. Given a JORDAN arc C with end-points A^*, B^*, we say that the equations

$$x = x(t), \quad y = y(t), \quad z = z(t), \quad a \leq t \leq b$$

define a monotonic transformation of the interval $a \leq t \leq b$ into a set on C, if the following conditions are satisfied.

1. $t = a$ and $t = b$ are carried into A^* and B^* respectively.
2. If $a < t_1 < t_2 < b$, then the point P_1, corresponding to t_1, is on the arc bounded by A^* and the image P_2 of t_2, end-points included (if P_2 coincides with A^*, this implies that P_1 also coincides with A^*).

Instead of an interval and a JORDAN arc, we can consider two JORDAN arcs C_1, C_2 and speak of a monotonic transformation of C_1 into a set on C_2, the meaning being too obvious to be explained.

If C_1 and C_2 both reduce to straight segments, say to the intervals $a \leq t \leq b$ and $\alpha \leq \tau \leq \beta$, then the conditions 1., 2. clearly mean that $\tau(t)$ is a monotonic function such that $\tau(a) = \alpha$, $\tau(b) = \beta$ or $\tau(a) = \beta$, $\tau(b) = \alpha$. This situation suggests a number of simple theorems concerning monotonic transformations which generalize theorems on monotonic functions. An important theorem on monotonic functions is the theorem of HELLY, to the effect that a uniformly bounded sequence of monotonic functions always contains an everywhere convergent subsequence, the limit function being obviously monotonic again. Before we state the generalization, we extend the notion of monotonic transformations to JORDAN curves. Given a JORDAN curve Γ^*, and a set of equations

$$x = x(\Theta), \quad y = y(\Theta), \quad z = z(\Theta), \quad 0 \leq \Theta < 2\pi,$$

we say that these equations define a monotonic transformation of the unit circle $u = \cos \Theta$, $v = \sin \Theta$ into a set on Γ^*, if the following conditions are satisfied.

1. There exist three distinct points A, B, C on the unit circle which are carried into three distinct points A^*, B^*, C^* of Γ^*.

2. The three non-overlapping arcs AB, BC, CA of the unit circle are transformed in a monotonic way into sets on the three non-overlapping arcs A^*B^*, B^*C^*, C^*A^* of Γ^*.

I.22. The theorem of HELLY admits then of the following obvious generalization. Let there be given a sequence Γ_n^* of JORDAN curves and for every one of them a monotonic transformation

$$T_n: x = x_n(\Theta), \qquad y = y_n(\Theta), \qquad z = z_n(\Theta)$$

of the unit circle $u = \cos\Theta, v = \sin\Theta$ into a set on Γ_n^*, such that three fixed distinct points A, B, C on the unit circle are carried into three distinct points A_n^*, B_n^*, C_n^* of Γ_n^*. Suppose the sequence Γ_n^* converges, in the FRÉCHET sense, to a JORDAN curve Γ^*, and also suppose that A_n^*, B_n^*, C_n^* converge to three distinct points A^*, B^*, C^* on Γ^*. Then it is possible to pick out a subsequence T_{n_k} such that the sequences $x_{n_k}(\Theta), y_{n_k}(\Theta), z_{n_k}(\Theta)$ converge everywhere on the unit circle. If the limit functions are denoted by $x(\Theta), y(\Theta), z(\Theta)$, then the equations $x = x(\Theta), y = y(\Theta), z = z(\Theta)$ define a monotonic transformation of the unit circle into a set on the limit curve Γ^*.

I.23. Let the equations

$$x = x(\Theta), \qquad y = y(\Theta), \qquad z = z(\Theta),$$

define a monotonic transformation T of the unit circle $u = \cos\Theta$, $v = \sin\Theta$ into a set on a JORDAN curve Γ^*. On account of the analogy with monotonic functions, there result the following facts for the functions $x(\Theta), y(\Theta), z(\Theta)$.

1. If Θ approaches from one side, that is to say in the counterclockwise or in the clockwise sense, any value Θ_0, then $x(\Theta), y(\Theta), z(\Theta)$ approach definite limits x_0^+, y_0^+, z_0^+ and x_0^-, y_0^-, z_0^- respectively.

2. If, for a certain Θ_0, we have $x_0^+ = x_0^-, y_0^+ = y_0^-, z_0^+ = z_0^-$, then $x(\Theta_0) = x_0^+ = x_0^-, y(\Theta_0) = y_0^+ = y_0^-, z(\Theta_0) = z_0^+ = z_0^-$, that is to say $x(\Theta), y(\Theta), z(\Theta)$ are then continuous for $\Theta = \Theta_0$.

3. If for two distinct points Θ_1, Θ_2 of the unit circle we have $x(\Theta_1) = x(\Theta_2), y(\Theta_1) = y(\Theta_2), z(\Theta_1) = z(\Theta_2)$, then $x(\Theta), y(\Theta), z(\Theta)$ all three reduce to constants on one of the two arcs determined by Θ_1 and Θ_2 on the unit circle.

I.24. While, in what precedes, we referred to the analogy with monotonic functions[1], J. DOUGLAS refers to results of FRÉCHET on parametric representations of the unit circle upon itself, to which the general case can be reduced. With every representation of the unit circle upon itself there can be associated a curve on a torus, and J. DOUGLAS obtains the necessary facts by a discussion of the situation on the torus[2].

---

[1] This point of view has been stressed by T. RADÓ: The problem of the least area and the problem of PLATEAU. Math. Z. Vol. 33 (1930) pp. 763—796

[2] See J DOUGLAS: Solution of the problem of PLATEAU. Trans. Amer. Math. Soc. Vol. 33 (1931) pp. 270—272.

Chapter II.

Minimal surfaces in the small.

II.1. Given a surface

$$S: x = x(u, v), \quad y = y(u, v), \quad z = z(u, v),$$

it is convenient to use vector notation and write simply $\mathfrak{x} = \mathfrak{x}(u, v)$, where $\mathfrak{x}(u, v)$ denotes the vector with components $x(u, v), y(u, v)$, $z(u, v)$. We have then for the first, second and third fundamental quantities of the surface the formulas[1]

$$E = \mathfrak{x}_u^2, \qquad F = \mathfrak{x}_u \mathfrak{x}_v, \qquad G = \mathfrak{x}_v^2,$$
$$L = \xi \mathfrak{x}_{uu}, \qquad M = \xi \mathfrak{x}_{uv}, \qquad N = \xi \mathfrak{x}_{vv},$$
$$e = \xi_u^2, \qquad f = \xi_u \xi_v, \qquad g = \xi_v^2.$$

ξ denotes the unit normal vector of the surface, the components X, Y, Z of which are given by the formulas

$$X = \frac{1}{W} \frac{\partial(y, z)}{\partial(u, v)}, \quad Y = \frac{1}{W} \frac{\partial(z, x)}{\partial(u, v)}, \quad Z = \frac{1}{W} \frac{\partial(x, y)}{\partial(u, v)},$$

where

$$W = (EG - F^2)^{\frac{1}{2}}.$$

For the total curvature K and for the mean curvature H respectively we have the formulas

$$K = \frac{LN - M^2}{W^2}, \quad H = \frac{EN - 2FM + GL}{2W^2}.$$

II.2. In writing all these formulas, we are making the obvious assumptions that $\mathfrak{x}(u, u)$ has the necessary differential coefficients and that $W > 0$. In differential geometry it is usually assumed that $\mathfrak{x}(u, v)$ is analytic. Many important issues are dodged by this latter blanket provision, particularly in the case of minimal surfaces in which we are interested. On the other hand, the assumption $W > 0$ is absolutely essential in differential geometry, since W appears in the denominator of practically every important formula.

It is convenient for our purposes to use the following definition. If $\mathfrak{x}(u, v)$ has continuous partial derivatives of the first order, and if $W > 0$, then the surface with the equation $\mathfrak{x} = \mathfrak{x}(u, v)$ will be called a regular surface of class C'. The classes $C'', C''', \ldots$ are defined in a similar way. The term *regular* refers to the condition $W > 0$. This condition secures the existence of the tangent plane. If we have occasion to change to new parameters, it is understood that we do this subject to the restriction that $\mathfrak{x}$ as function of the new parameters satisfies the same conditions.

---

[1] See for instance BLASCHKE: Vorlesungen über Differentialgeometrie. Vol. 1.

If $H \equiv 0$ for a regular surface S of class C'', then S is called a *minimal surface*. The term is misleading inasmuch as the area of a minimal surface, bounded by a given curve, is generally not a minimum (see III.14). A surface S, given by an equation $z = z(x, y)$, is a minimal surface if and only if $z(x, y)$ satisfies the partial differential equation

$$(1 + q^2)r - 2pqs + (1 + p^2)t = 0.$$

Indeed, the left-hand side is the numerator of the expression to which the general formula for H reduces for a surface given in the special representation $z = z(x, y)$.

II.3. It will be convenient to use the so-called differential notation in the sequel. If φ, ψ, ω are functions of u, v in a simply connected domain D, then the equation

$$d\omega = \varphi \, du + \psi \, dv$$

means that $\omega_u = \varphi$, $\omega_v = \psi$. We shall say then that $\varphi \, du + \psi \, dv$ is a complete differential in D, or that $\varphi \, du + \psi \, dv$ is the differential of ω. If φ, ψ have continuous partial derivatives of the first order, then $\varphi \, du + \psi \, dv$ is a complete differential if and only if $\varphi_v = \psi_u$.

If $\omega, \varphi_1, \varphi_2, \psi_1, \psi_2$ are functions of u, v in D, then the equation

$$d\omega = \varphi_1 \, d\varphi_2 + \psi_1 \, d\psi_2 \qquad (2.1)$$

means that on substituting

$$d\varphi_2 = \varphi_{2u} \, du + \varphi_{2v} \, dv, \qquad d\psi_2 = \psi_{2u} \, du + \psi_{2v} \, dv,$$

the right-hand side becomes the differential of ω in the sense explained above.

The convenience of the differential notation results from the fact that if new variables α, β are introduced by equations $\alpha = \alpha(u, v)$, $\beta = \beta(u, v)$, then $\omega, \varphi_1, \varphi_2, \psi_1, \psi_2$ considered as functions of α, β also satisfy the equation (2.1).

II.4. The study of minimal surfaces originated with the problem of the least area, that is to say the problem of determining and investigating a surface, bounded by a given curve, the area of which is a minimum[1].

Let $\varGamma^*$ be the given boundary curve, and suppose there exists a surface S, bounded by $\varGamma^*$, the area of which is a minimum. Suppose also that S is a regular surface of class C''[2]. We ask then for consequences of the minimizing property of S, that is to say for necessary conditions for a minimum area. We are going to review briefly several of the ways in which this question has been handled in the older literature, restricting ourselves to cases which are significant from the point of view of the recent investigations.

---

[1] A beautiful presentation of the theory, on the basis of the older literature, is given in the classical work of DARBOUX: Théorie générale des surfaces. Vol. 1.

[2] The usual text-book assumptions are even more restrictive and therefore even less natural. Cf. II.12.

II.5. Choose a point P_0 on the minimizing surface S. If the axes x, y, z are properly chosen, then a sufficiently small vicinity S_0 of P_0 admits of a representation of the form $z = z(x, y)$, where $z(x, y)$ is single-valued in a certain region R_0 of the xy-plane and has continuous partial derivatives of the first and second order in R_0. From the minimizing property of S it follows that S_0 has also the minimizing property with respect to its own boundary curve. The area $\mathfrak{A}(S_0)$ of S_0 is given by

$$\mathfrak{A}(S_0) = \iint\limits_{R_0} (1 + p^2 + q^2)^{\frac{1}{2}} \, dx \, dy \,.$$

Thus $z(x, y)$ minimizes this integral with respect to functions which coincide with $z(x, y)$ on the boundary of R_0.

Denote by $\lambda(x, y)$ a function which vanishes on the boundary of R_0 and has continuous derivatives of the first and second order in R_0. Let ε be a parameter. Then the area of the surface $z = z(x, y) + \varepsilon \lambda(x, y)$ is a function $A(\varepsilon)$ of ε which has a minimum for $\varepsilon = 0$. Hence

$$A'(0) = \iint\limits_{R_0} \left(\frac{p}{W} \lambda_x + \frac{q}{W} \lambda_y\right) dx \, dy = 0 \,, \tag{2.2}$$

where

$$W = (1 + p^2 + q^2)^{\frac{1}{2}}.$$

Since (2.2) holds for all functions $\lambda(x, y)$ with the properties described above, it follows[1] that

$$\frac{\partial}{\partial x} \frac{p}{W} + \frac{\partial}{\partial y} \frac{q}{W} = 0 \,, \tag{2.3}$$

or

$$(1 + q^2)r - 2pqs + (1 + p^2)t = 0.$$

Hence (see II.2) the mean curvature of S_0 vanishes identically. As S_0 is the vicinity of an arbitrary point of S, it follows that the minimizing surface is a minimal surface.

II.6. The variation problem

$$\iint (1 + p^2 + q^2)^{\frac{1}{2}} \, dx \, dy = \text{minimum}$$

was the example which LAGRANGE considered to illustrate his method, yielding the so-called EULER-LAGRANGE equation, for the case of multiple integrals. As observed by LAGRANGE[2], the equation (2.3) means that

$$\frac{p \, dy - q \, dx}{(1 + p^2 + q^2)^{\frac{1}{2}}}$$

is a complete differential (cf. II.3).

II.7. Let again S be the minimizing surface, given in a general representation $\mathfrak{x} = \mathfrak{x}(u, v)$, where (u, v) varies in a region R. Let $\lambda(u, v)$ be a function which vanishes on the boundary of R and has

---

[1] See for instance BOLZA: Vorlesungen über Variationsrechnung, pp. 653—655.
[2] See DARBOUX: Théorie générale des surfaces. Vol. 1 pp. 267—268.

continuous derivatives of the first order in R, and let ε be a parameter. The area of the surface[1]

$$\mathfrak{x} = \mathfrak{x}(u, v) + \varepsilon \lambda(u, v)\, \xi(u, v),$$

where $\xi(u, v)$ is the unit normal vector of S, is then a function $A(\varepsilon)$ of ε which has a minimum for $\varepsilon = 0$. Hence $A'(0) = 0$, $A''(0) \geqq 0$. The first condition gives

$$\iint\limits_{R} HW\lambda\, du\, dv = 0.$$

As this condition holds for every function λ with the properties specified above, it follows again that $H \equiv 0$, that is to say that S is a minimal surface. The condition $A''(0) \geqq 0$ gives then that

$$\iint\limits_{R} \frac{1}{W}\{\lambda^2[Eg-2Ff+Ge+4(LN-M^2)]+E\lambda_v^2-2F\lambda_u\lambda_v+G\lambda_u^2\}du\,dv\geqq 0, \quad (2.4)$$

for all functions λ with the properties specified above. This condition permits us to construct examples of minimal surfaces which do not have the minimizing property (see III.14).

II.8. Let again the minimizing surface S be given in a general representation

$$S:x = x(u, v), \quad y = y(u, v), \quad z = z(u, v), \quad (u, v) \text{ in } R.$$

Let $\lambda(u, v)$ and ε have the same meaning as in II.7. Then the area of the surface

$$x = x(u, v) + \varepsilon\lambda(u, v), \quad y = y(u, v), \quad z = z(u, v)$$

is a function $A(\varepsilon)$ of ε which has a minimum for $\varepsilon = 0$. The condition $A'(0) = 0$ gives this time the equation[2]

$$\frac{\partial}{\partial u}\frac{Gx_u - Fx_v}{W} + \frac{\partial}{\partial v}\frac{Ex_v - Fx_u}{W} = 0. \quad (2.5)$$

Similar variations of the y and z coordinates respectively yield the further equations

$$\frac{\partial}{\partial u}\frac{Gy_u - Fy_v}{W} + \frac{\partial}{\partial v}\frac{Ey_v - Fy_u}{W} = 0, \quad (2.6)$$

$$\frac{\partial}{\partial u}\frac{Gz_u - Fz_v}{W} + \frac{\partial}{\partial v}\frac{Ez_v - Fz_u}{W} = 0. \quad (2.7)$$

In other words, the three expressions

$$\frac{Ex_v - Fx_u}{W}du - \frac{Gx_u - Fx_v}{W}dv, \quad (2.8)$$

$$\frac{Ey_v - Fy_u}{W}du - \frac{Gy_u - Fy_v}{W}dv, \quad (2.9)$$

$$\frac{Ez_v - Fz_u}{W}du - \frac{Gz_u - Fz_v}{W}dv \quad (2.10)$$

are complete differentials.

---

[1] See Darboux: Théorie générale des surfaces. Vol. 1 pp. 281—284.

[2] For the following formulas, see Bolza: Vorlesungen über Variationsrechnung, p. 667.

II.9. The preceding results have important implications. Suppose that the minimizing surface S is given in terms of isothermic parameters[1]. Then $E = G$, $F = 0$, and the equations (2.5), (2.6), (2.7) reduce to

$$x_{uu} + x_{vv} = 0, \qquad y_{uu} + y_{vv} = 0, \qquad z_{uu} + z_{vv} = 0.$$

That is to say: *if the minimizing surface S is given in terms of isothermic parameters, then the coordinate functions $x(u, v)$, $y(u, v)$, $z(u, v)$ are harmonic functions.*

II.10. The preceding result may be obtained also by using the fact that a harmonic function $h(u, v)$ with given boundary values minimizes the DIRICHLET integral

$$\iint (h_u^2 + h_v^2)\, du\, dv.$$

Conversely, if a function $h(u, v)$ has this minimizing property, then $h(u, v)$ is harmonic. Consider then[2] the minimizing surface

$$S: \mathfrak{x} = \mathfrak{x}(u, v), \qquad (u, v) \text{ in } R,$$

given in terms of isothermic parameters. On account of $E = G, F = 0$, we have then for the area $\mathfrak{A}(S)$ of S:

$$\mathfrak{A}(S) = \iint (EG - F^2)^{\frac{1}{2}} = \iint E = \iint G = \tfrac{1}{2} \iint (E + G).$$

Denote by $\bar{\mathfrak{x}}(u, v)$ the vector whose components $\bar{x}(u, v), \bar{y}(u, v), \bar{z}(u, v)$ are the harmonic functions coinciding with the components $x(u, v)$, $y(u, v), z(u, v)$ of $\mathfrak{x}(u, v)$ on the boundary of R, and denote by $\bar{S}$ the surface $\bar{S}: \mathfrak{x} = \bar{\mathfrak{x}}(u, v)$, (u, v) in R. On account of the minimizing property of harmonic functions we have then

$$\iint (\bar{x}_u^2 + \bar{x}_v^2) \leqq \iint (x_u^2 + x_v^2), \tag{2.11}$$

$$\iint (\bar{y}_u^2 + \bar{y}_v^2) \leqq \iint (y_u^2 + y_v^2), \tag{2.12}$$

$$\iint (\bar{z}_u^2 + \bar{z}_v^2) \leqq \iint (z_u^2 + z_v^2), \tag{2.13}$$

and there follows by addition the inequality

$$\iint (\bar{E} + \bar{G}) \leqq \iint (E + G).$$

On account of the minimizing property of S, we have $\mathfrak{A}(S) \leqq \mathfrak{A}(\bar{S})$. Summing up, we can write:

$$\mathfrak{A}(S) \leqq \mathfrak{A}(\bar{S}) = \iint (\bar{E}\bar{G} - \bar{F}^2)^{\frac{1}{2}} \leqq \iint \bar{E}^{\frac{1}{2}}\bar{G}^{\frac{1}{2}} \leqq \tfrac{1}{2} \iint (\bar{E} + \bar{G}) \leqq \tfrac{1}{2} \iint (E + G) = \mathfrak{A}(S).$$

Consequently we must have the sign of equality all over, which implies the sign of equality in (2.11), (2.12), (2.13). That is to say, $x(u, v)$,

---

[1] See for instance BOLZA: Vorlesungen über Variationsrechnung, p. 667.

[2] The following reasoning is both a special case ($\sigma = 0$) and the origin of that used in V.20.

$y(u, v), z(u, v)$ minimize the DIRICHLET integral, and they are consequently harmonic functions.

II.11. Suppose the minimizing surface S is given by an equation $z = z(x, y)$. The expressions (2.8), (2.9), (2.10) reduce then to

$$-\frac{pq}{W} dx - \frac{1+q^2}{W} dy, \quad \frac{1+p^2}{W} dx + \frac{pq}{W} dy, \quad \frac{q}{W} dx - \frac{p}{W} dy, \quad (2.14)$$

where $W = (1 + p^2 + q^2)^{\frac{1}{2}}$. The components X, Y, Z of the unit normal vector are given by

$$X = -\frac{p}{W}, \quad Y = -\frac{q}{W}, \quad Z = \frac{1}{W}.$$

The expressions (2.14) are then found to be identical to

$$Y dz - Z dy, \quad Z dx - X dz, \quad X dy - Y dx.$$

Hence: for the minimizing surface S the expressions

$$Y dz - Z dy, \quad Z dx - X dz, \quad X dy - Y dx$$

are complete differentials[1].

II.12. The preceding properties of the minimizing surface S have been obtained under very restrictive assumptions concerning S. We shall see (III.13) that without proper restrictions concerning S the conclusion that S is a minimal surface is not in general valid.

There arises therefore the problem of determining the extent to which the restrictions imposed upon S can be lessened without hurting the validity of the conclusions obtained in the preceding sections. The discussion of this problem constitutes an essential part of most of the methods developed to handle the existence theorems in the problem of PLATEAU. The generality of the conclusions obtained in II.5, II.6, II.9 and II.11 will be considered in Chapter IV. The reasoning of II.10 will be generalized in Chapter V and Chapter VI in two different ways.

II.13. Since a minimal surface (as defined in II.2) generally does not have a minimum area, the question arises as to whether the conclusions obtained in the preceding sections for surfaces with a minimum area do or do not remain valid for minimal surfaces regardless of whether or not they have the minimizing property. This question leads in a very natural manner to the body of those theorems in the theory of minimal surfaces which are important for a clear understanding of the investigations reviewed in this report, and therefore we are going to discuss briefly that question.

II.14. *A regular surface S of class C'' is a minimal surface if and only if $Y dz - Z dy, Z dx - X dz, X dy - Y dx$ are complete differentials $(X, Y, Z$ are the components of the unit normal vector of the surface).*

This theorem of H. A. SCHWARZ may be proved as follows. It clearly is sufficient to verify the theorem for small portions of S. Let

---

[1] Cf. II 14

S_0 be a small simply connected portion of S. Then S_0 can be represented in one of the forms $z = z(x, y)$, $x = x(y, z)$, $y = y(z, x)$, say in the form $z = z(x, y)$. Then

$$X = -\frac{p}{W}, \quad Y = -\frac{q}{W}, \quad Z = \frac{1}{W}, \quad W = (1 + p^2 + q^2)^{\frac{1}{2}},$$

and

$$Y\,dz - Z\,dy = -\frac{pq}{W}\,dx - \frac{1+q^2}{W}\,dy, \tag{2.15}$$

$$Z\,dx - X\,dz = \frac{1+p^2}{W}\,dx + \frac{pq}{W}\,dy, \tag{2.16}$$

$$X\,dy - Y\,dx = \frac{q}{W}\,dx - \frac{p}{W}\,dy. \tag{2.17}$$

To apply the cross-wise differentiation test, we compute

$$\frac{\partial}{\partial y}\left(-\frac{pq}{W}\right) - \frac{\partial}{\partial x}\left(-\frac{1+q^2}{W}\right) = -\frac{p}{W^3}\,T,$$

$$\frac{\partial}{\partial y}\left(\frac{1+p^2}{W}\right) - \frac{\partial}{\partial x}\left(\frac{pq}{W}\right) = -\frac{q}{W^3}\,T,$$

$$\frac{\partial}{\partial y}\left(\frac{q}{W}\right) - \frac{\partial}{\partial x}\left(-\frac{p}{W}\right) = \frac{1}{W^3}\,T,$$

where

$$T = (1 + q^2)r - 2pqs + (1 + p^2)t.$$

These formulas show that the expressions (2.15), (2.16), (2.17) are complete differentials if and only if $T \equiv 0$, that is to say if and only if the surface is a minimal surface.

II.15. *If S is a minimal surface, and S_0 is a sufficiently small simply connected vicinity of an interior point P_0 of S, then S_0 can be mapped upon a plane region in a one-to-one and conformal way.*

It should be observed that since S is supposed to be of class C'', the theorem of I.18 is sufficiently general to apply. While reference to that theorem implies the use of the involved arguments necessary for its proof, the fact that S is a minimal surface enables us to prove in a direct and elementary way the existence of a conformal map[1].

The axes x, y, z being properly chosen, a sufficiently small simply connected vicinity S_0 of an interior point P_0 of S admits of a representation $z = z(x, y)$, where $z(x, y)$ is single-valued in a certain domain D_0 of the xy-plane and has there continuous partial derivatives of the first and second order. It is legitimate to suppose that D_0 is the interior of a circle. Since S_0 is a minimal surface, the expressions (2.15), (2.16), (2.17) in II.14 are complete differentials:

$$-\frac{pq}{W}\,dx - \frac{1+q^2}{W}\,dy = d\omega_1,$$

$$\frac{1+p^2}{W}\,dx + \frac{pq}{W}\,dy = d\omega_2,$$

$$\frac{q}{W}\,dx - \frac{p}{W}\,dy = d\omega_3,$$

---

[1] Ch. H. Müntz: Die Lösung des Plateauschen Problems über konvexen Bereichen. Math. Ann Vol. 94 (1925) pp. 53—96. — T. Radó: Über den analytischen Charakter der Minimalflächen. Math. Z. Vol. 24 (1925) pp. 321—327.

where ω_1, ω_2, ω_3 are single-valued functions in D_0. Introduce then new variables α, β by the equations $\alpha = x$, $\beta = \omega_1(x, y)$. Since

$$\frac{\partial \beta}{\partial y} = -\frac{1 + q^2}{W} < 0,$$

it is readily seen that D_0 is carried in a one-to-one and continuous way into a certain domain D_0^* of the $\alpha\beta$-plane. We obtain then the formulas

$$dx = d\alpha, \tag{2.18}$$

$$dy = -\frac{pq}{1 + q^2}\, d\alpha - \frac{W}{1 + q^2}\, d\beta, \tag{2.19}$$

$$dz = \frac{p}{1 + q^2}\, d\alpha - \frac{qW}{1 + q^2}\, d\beta, \tag{2.20}$$

$$d\omega_1 = d\beta, \tag{2.21}$$

$$d\omega_2 = \frac{W}{1 + q^2}\, d\alpha - \frac{pq}{1 + q^2}\, d\beta, \tag{2.22}$$

$$d\omega_3 = \frac{qW}{1 + q^2}\, d\alpha + \frac{p}{1 + q^2}\, d\beta. \tag{2.23}$$

The first three formulas yield $x_\alpha, x_\beta, y_\alpha, y_\beta, z_\alpha, z_\beta$, and direct computation shows that

$$x_\alpha^2 + y_\alpha^2 + z_\alpha^2 = \frac{1 + p^2 + q^2}{1 + q^2} = x_\beta^2 + y_\beta^2 + z_\beta^2,$$

$$x_\alpha x_\beta + y_\alpha y_\beta + z_\alpha z_\beta = 0,$$

that is to say that α, β are isothermic parameters.

The lines $\alpha = $ constant of this conformal map correspond to the lines $x = $ constant on the surface. Hence the preceding computations prove the important theorem, discovered by RIEMANN and by BELTRAMI, that a minimal surface is intersected by parallel planes in curves which constitute an isothermic family.

II.16. The formulas (2.18) to (2.23) in II.15 show that

$$x_\alpha = \omega_1{}_\beta, \qquad x_\beta = -\omega_1{}_\alpha,$$

$$y_\alpha = \omega_2{}_\beta, \qquad y_\beta = -\omega_2{}_\alpha,$$

$$z_\alpha = \omega_3{}_\beta, \qquad z_\beta = -\omega_3{}_\alpha.$$

That is to say, x and ω_1, y and ω_2, z and ω_3 satisfy, as functions of α and β, the CAUCHY-RIEMANN equations. It follows by inspection of the formulas defining these functions that they have continuous partial derivatives of the first order. Consequently, x, y, z as functions of α and β are harmonic and therefore analytic functions of α, β. Hence: *minimal surfaces are analytic*[1], although their definition (see II.2) implies only that they are of class C''. It follows, in particular, that

---

[1] CH. H. MÜNTZ: Die Lösung des PLATEAUschen Problems über konvexen Bereichen. Math. Ann. Vol. 94 (1925) pp. 53—96. — T. RADÓ: Über den analytischen Charakter der Minimalflächen. Math. Z. Vol. 24 (1925) pp. 321—327.

every solution $z(x, y)$, with continuous partial derivatives of the first and second order, of the equation

$$(1 + q^2)r - 2pqs + (1 + p^2)t = 0$$

is analytic.

II.17. The following theorem of WEIERSTRASS is fundamental for the theory of minimal surfaces.

Given a regular surface $S : \mathfrak{x} = \mathfrak{x}(u, v)$ *of class* C'' *in terms of iso-thermic parameters. Then* S *is a minimal surface if and only if the components* $x(u, v), y(u, v), z(u, v)$ *of* $\mathfrak{x}(u, v)$ *are harmonic functions.*

To prove this, show first that on account of $E = G, F = 0$ we have the identities:

$$Y dz - Z dy = x_v du - x_u dv,$$

$$Z dx - X dz = y_v du - y_u dv,$$

$$X dy - Y dx = z_v du - z_u dv.$$

Using the cross-wise differentiation test, we see that the expressions on the right-hand sides are complete differentials if and only if x, y, z are harmonic. On account of the theorem of SCHWARZ (see II.14), the expressions on the left-hand sides are complete differentials if and only if the surface is minimal. The theorem of WEIERSTRASS is an immediate consequence of these two facts.

II.18. The theorem of WEIERSTRASS leads to *standard formulas for minimal surfaces* which are fundamental for the existence theorems in the problem of PLATEAU.

According to II.15, a minimal surface admits in the small of iso-thermic parameters. That is to say, every minimal surface can be represented, in the small, by equations

$$x = x(u, v), \quad y = y(u, v), \quad z = z(u, v), \quad (u, v) \text{ in } D,$$

with $E = G, F = 0$. We can suppose that D is the interior of a circle. According to the theorem of WEIERSTRASS, $x(u, v), y(u, v), z(u, v)$ are then harmonic functions. We can write therefore:

$$x = \Re f_1(w), \quad y = \Re f_2(w), \quad z = \Re f_3(w),$$

where f_1, f_2, f_3 are analytic functions of the complex variable $w = u + iv$. It follows then that

$$x_u - i x_v = \varphi_1, \quad y_u - i y_v = \varphi_2, \quad z_u - i z_v = \varphi_3, \quad (2.24)$$

where $\varphi_1 = f_1', \varphi_2 = f_2', \varphi_3 = f_3'$ are again analytic functions of w. The condition $EG - F^2 > 0$, expressing that the surface is regular, is found to be equivalent to the condition that $\varphi_1, \varphi_2, \varphi_3$ do not vanish simultaneously at any point in D. Squaring and adding, we obtain from (2.24) the equation

$$E - G - 2iF = \varphi_1^2 + \varphi_2^2 + \varphi_3^2.$$

Thus $E = G, F = 0$ is equivalent to $\varphi_1^2 + \varphi_2^2 + \varphi_3^2 = 0$. This gives the theorem (discovered by MONGE):

Every minimal surface can be represented in the small by equations of the form

$$x = \Re \int^w \varphi_1 \, dw \,, \qquad y = \Re \int^w \varphi_2 \, dw \,, \qquad z = \Re \int^w \varphi_3 \, dw \,, \qquad (2.25)$$

where φ_1, φ_2, φ_3 are analytic functions of w in the interior D of some circle and satisfying there the two conditions:

1. $\varphi_1^2 + \varphi_2^2 + \varphi_3^2 = 0$ *in D.*
2. $\varphi_1, \varphi_2, \varphi_3$ *do not vanish simultaneously at any point of D.*

Conversely, if $\varphi_1, \varphi_2, \varphi_3$ satisfy these conditions, then the equations (2.25) define a minimal surface.

II.19. The determination of the totality of minimal surfaces reduces therefore to the determination of all triples of functions $\varphi_1(w)$, $\varphi_2(w)$, $\varphi_3(w)$ with the properties 1. and 2. This can be done in several ways.

Choose a point w_0 in D. Since $\varphi_1(w_0)$, $\varphi_2(w_0)$, $\varphi_3(w_0)$ are not all three equal to zero, suppose that $\varphi_3(w_0) \neq 0$, for instance. We restrict ourselves to a vicinity D_0 of w_0, where $\varphi_3 \neq 0$. From $\varphi_1^2 + \varphi_2^2 + \varphi_3^2 = 0$ it follows that

$$(\varphi_1 - i\varphi_2)(\varphi_1 + i\varphi_2) = -\varphi_3^2. \qquad (2.26)$$

Consequently $\varphi_1 - i\varphi_2 \neq 0$ in D_0. Put

$$\mu = \frac{\varphi_1 - i\varphi_2}{2}, \qquad \lambda = \frac{\varphi_3}{2\mu}. \qquad (2.27)$$

Then λ, μ are analytic functions of w in D_0, and $\mu \neq 0$, $\lambda \neq 0$ in D_0. From the equations (2.26), (2.27) it follows then that

$$\varphi_1 = (1 - \lambda^2)\mu \,,$$
$$\varphi_2 = i(1 + \lambda^2)\mu \,,$$
$$\varphi_3 = 2\lambda\mu \,.$$

Conversely, if λ, μ are analytic functions of w, then the functions φ_1, φ_2, φ_3 satisfy the condition $\varphi_1^2 + \varphi_2^2 + \varphi_3^2 = 0$, as follows by direct computation. The formulas also show that if $\mu \neq 0$ in a domain D_0, then $\varphi_1, \varphi_2, \varphi_3$ do not vanish simultaneously at any point of D_0. Hence: the equations

$$x = \Re \int^w (1 - \lambda^2)\mu \, dw \,,$$
$$y = \Re \int^w i(1 + \lambda^2)\mu \, dw \,, \qquad (2.28)$$
$$z = \Re \int^w 2\lambda\mu \, dw \,,$$

where λ, μ are analytic functions of w in the interior D_0 of some circle and $\mu \neq 0$ in D_0, yield the totality of minimal surfaces, considered in the small.

II.20. In II.19, we supposed that $\varphi_3 \neq 0$ for instance. This situation always can be arranged for by renaming the axes x, y, z, if necessary. It should be observed that it might happen that it is absolutely necessary to rename the axes in order to represent a given minimal surface by the formulas (2.28). Consider for instance the equations

$$x = \Re \int^{w} \varphi_1 \, dw, \qquad y = \Re \int^{w} \varphi_2 \, dw, \qquad z = \Re \int^{w} \varphi_3 \, dw,$$

where $\varphi_1 = w^2 - 1$, $\varphi_2 = i(1 + w^2)$, $\varphi_3 = 2w$, in the vicinity of $w = 0$. Then $\varphi_1^2 + \varphi_2^2 + \varphi_3^2 = 0$, and φ_1, φ_2, φ_3 never vanish simultaneously. Still, φ_1, φ_2, φ_3 cannot be represented in the form (2.28) in the vicinity of $w = 0$. Indeed, from the equations

$$w^2 - 1 = (1 - \lambda^2)\mu, \qquad i(1 + w^2) = i(1 + \lambda^2)\mu, \qquad 2w = 2\lambda\mu$$

it would follow that $\mu = w^2$, $\lambda = \dfrac{1}{w}$. Thus λ would necessarily have a pole at $w = 0$, and μ would necessarily vanish at $w = 0$.

Thus the statement at the end of II.19 should be amended by saying that the formulas (2.28) yield all minimal surfaces provided we rename, if necessary, the axes x, y, z. This constitutes a very serious disadvantage of the formulas (2.28)[1].

II.21. A somewhat closer discussion shows that every triple φ_1, φ_2, φ_3 with the properties 1., 2. stated in II.18 can be represented, in the small, by the formulas

$$\begin{aligned}
\varphi_1 &= \Phi^2 - \Psi^2, \\
\varphi_2 &= i(\Phi^2 + \Psi^2), \\
\varphi_3 &= 2\Phi\Psi,
\end{aligned}$$

where Φ, Ψ are single-valued analytic functions of w which do not have any common zeros. This gives the *formulas of* WEIERSTRASS: every minimal surface can be represented, in the small, by equations

$$\left. \begin{aligned}
x &= \Re \int^{w} (\Phi^2 - \Psi^2) \, dw, \\
y &= \Re \int^{w} i(\Phi^2 + \Psi^2) \, dw, \\
z &= \Re \int^{w} 2\Phi\Psi \, dw,
\end{aligned} \right\} \qquad (2.29)$$

where Φ, Ψ are single-valued analytic functions of w without common zeros[2]. The converse is obvious.

II.22. In later Chapters we shall have to consider minimal surfaces in a more general sense. The generalization will consist in dropping the

---

[1] A large part of the older work is based on the special case of (2.28) corresponding to $\lambda(w) = w$.

[2] The classical notation is G, H. As G, H have a standard meaning in differential geometry, we changed to Φ, Ψ.

condition of regularity $EG - F^2 > 0$ of the surface. On account of (2.24) this amounts to permitting the functions $\varphi_1, \varphi_2, \varphi_3$ of II.18 to vanish simultaneously. The question arises then if the formulas (2.29) still represent, in the small, all minimal surfaces. This is not the case, as might be inferred from the example

$$x = \Re \int^w \varphi_1 \, dw, \qquad y = \Re \int^w \varphi_2 \, dw, \qquad z = \Re \int^w \varphi_3 \, dw,$$

where $\varphi_1 = 3w$, $\varphi_2 = 5iw$, $\varphi_3 = 4w$, and the situation is considered in the vicinity of $w = 0$. Clearly $\varphi_1^2 + \varphi_2^2 + \varphi_3^2 = 0$. Suppose there would exist, in the vicinity of $w = 0$, two single-valued analytic functions Φ, Ψ such that $3w = \Phi^2 - \Psi^2$, $5iw = i(\Phi^2 + \Psi^2)$, $4w = 2\Phi\Psi$. It would follow that $\Phi^2 = 4w$, and it is then clear that Φ is not single-valued in the vicinity of $w = 0$.

An obvious remedy would be to permit the functions Φ, Ψ, in the formulas (2.29), to have algebraic singularities. As far as the investigations to be reviewed in the sequel are concerned, this step has not been taken (cf. Chapter V).

II.23. Since the purpose of this Chapter is to present only those facts about minimal surfaces which are significant from the point of view of the recent work on the problem of PLATEAU, we restrict ourselves to recall one more theorem, due to H. A. SCHWARZ[1]. *Suppose a minimal surface S contains a straight line g. Then g is an axis of symmetry of S.*

A simple proof may be obtained as follows. Choose the axes x, y, z in such a way that S may be represented, in the vicinity of one of its points situated on g, in the form $z = z(x, y)$, and that the y-axis coincides with g. Introduce then the isothermic parameters α, β defined in II.15. Since $z(0, y) \equiv 0$ and consequently $q(0, y) \equiv 0$ in the present case, it follows, with regard to the formulas in II.15, that

$$z = 0, \qquad x = 0, \qquad y_\alpha = 0 \text{ for } \alpha = 0. \tag{2.30}$$

Since x, y, z and consequently y_α, as functions of α and β, are harmonic functions (see II.16), it follows from (2.30) that

$$z(\alpha, \beta) = -z(-\alpha, \beta), \qquad x(\alpha, \beta) = -x(-\alpha, \beta),$$
$$y_\alpha(\alpha, \beta) = -y_\alpha(-\alpha, \beta)$$

on account of the principle of symmetry. From the last of these three equations it follows by integration that $y(\alpha, \beta) = y(-\alpha, \beta)$. The equations $x(\alpha, \beta) = -x(-\alpha, \beta)$, $y(\alpha, \beta) = y(-\alpha, \beta)$, $z(\alpha, \beta) = -z(-\alpha, \beta)$ show that if (x_0, y_0, z_0) is a point of our minimal surface S, then $(-x_0, y_0, -z_0)$ is also a point of S. This proves the theorem of SCHWARZ.

---

[1] Gesammelte Mathematische Abhandlungen Vol. 1 p. 181.

Chapter III.

Minimal surfaces in the large.

III.1. The problem of PLATEAU, as considered for instance by H. A. SCHWARZ in his classical investigations[1], calls for a minimal surface $\mathfrak{M}$ which is bounded by a given Jordan curve Γ^* and which is *free of singularities in its interior*. The question naturally arises if this supplementary condition can be complied with if Γ^* is very complicated. The very particular cases which have been dealt with in the older literature certainly cannot be considered as representative of what might be expected in the case of a general JORDAN curve. At any rate, that supplementary condition has been dropped, at least for the time being, in the modern general investigations, and the existence of a solution free of singularities has been established only for certain special classes of curves.

III.2. If the solution is permitted to have singularities, it clearly is necessary to specify the nature of these singularities. For instance, if the solution would be permitted to have *edges*, then every polyhedron, bounded by a given polygon Γ^*, should be considered as a solution of the problem of PLATEAU for the contour Γ^*. Or consider a uniqueness theorem, for instance the theorem stated by H. A. SCHWARZ that the solution of the problem of PLATEAU is unique if the given contour is a skew quadrilateral[2]. H. A. SCHWARZ had only solutions in mind which are free of singularities. We shall see (III.12) that the theorem remains valid even if the solution is permitted to have certain singularities. On the other hand, the theorem breaks down if the specifications as to permissible singularities are too liberal, as follows from the remark made above.

III.3. The specifications as to the permissible singularities are by no means uniform in the recent literature. In other words, the different authors do not always consider the same problem. We first consider the most liberal specifications actually used in the literature.

Given a continuous surface $S : \mathfrak{x} = \mathfrak{x}(u, v)$, (u, v) in R, where R denotes a JORDAN region bounded by a JORDAN curve Γ, and given also a JORDAN curve Γ^* in the xyz-space, we shall say that S is a *minimal surface (of the type of the circular disc) bounded by* Γ^* if the following conditions are satisfied[3].

1. The equations of S carry Γ in a topological way into Γ^*.

---

[1] Gesammelte Mathematische Abhandlungen Vol. 1.

[2] Gesammelte Mathematische Abhandlungen Vol. 1 p. 111.

[3] T. RADÓ: Contributions to the theory of minimal surfaces. Acta Litt. Sci. Szeged Vol. 6 (1932) pp. 1—20.

2. For every point (u_0, v_0) interior to Γ there exists a vicinity V_0 and a one-to-one and continuous transformation $\alpha = \alpha_0(u, v)$, $\beta = \beta_0(u, v)$ of V_0 into some region $\overline{V}_0$ of the $\alpha\beta$-plane such that the components x, y, z of $\mathfrak{x}$ as functions of α, β are harmonic in $\overline{V}_0$ and satisfy there the relations
$$\overline{E} = \overline{G}, \quad \overline{F} = 0,$$
where $\overline{E} = \mathfrak{x}_\alpha^2$, $\overline{F} = \mathfrak{x}_\alpha \mathfrak{x}_\beta$, $\overline{G} = \mathfrak{x}_\beta^2$. Parameters α, β with this property will then be called local typical parameters for the vicinity of (u_0, v_0).

III.4. If $\overline{E}\overline{G} - \overline{F}^2 > 0$ at the point (α_0, β_0) into which (u_0, v_0) is carried, then the corresponding point of S will be called a *regular point*. The vicinity, on S, of a regular point is a minimal surface in the sense of differential geometry.

If $\overline{E}\overline{G} - \overline{F}^2 = 0$ at (α_0, β_0), then we have, on account of $\overline{E} = \overline{G}$, $\overline{F} = 0$, $x_\alpha = y_\alpha = z_\alpha = x_\beta = y_\beta = z_\beta = 0$ at (α_0, β_0). Suppose that all the partial derivatives of x, y, z with respect to α, β vanish at (α_0, β_0) up to and including a certain order $n \geq 1$, while at least one of the derivatives of order $n + 1$ is different from zero. Then the point corresponding to (α_0, β_0) on S will be called a *branch-point* of order n.

The preceding notions are independent of the special choice of the local typical parameters α, β. If S has only regular points, then S as a whole is a minimal surface in the sense of differential geometry, and will be called a *regular minimal surface*.

III.5. We now shall state the problems which will be discussed in this report. First we have the following statements of the problem of PLATEAU, which have been considered actually in the literature.

Problem P_1. Given, in the xyz-space, a JORDAN curve Γ^*, determine a minimal surface, of the type of the circular disc, bounded by Γ^* (the term *minimal surface* being used in the sense of III.3).

Problem P_2. Solve problem P_1 under the supplementary condition that the solution admits of a representation $S : \mathfrak{x} = \mathfrak{x}(u, v)$, $u^2 + v^2 \leq 1$, where the components $x(u, v)$, $y(u, v)$, $z(u, v)$ of $\mathfrak{x}(u, v)$ are continuous for $u^2 + v^2 \leq 1$, harmonic for $u^2 + v^2 < 1$, and satisfy for $u^2 + v^2 < 1$ the equations $E = G$, $F = 0$. Furthermore, the equations $x = x(u, v)$, $y = y(u, v)$, $z = z(u, v)$ are required to carry $u^2 + v^2 = 1$ in a topological way into the given JORDAN curve Γ^*.

Problem P_3. Solve problem P_2 under the supplementary condition that the functions $x(u, v)$, $y(u, v)$, $z(u, v)$ admit of a representation of the form

$$x = \Re \int^w (\Phi^2 - \Psi^2)\, dw,$$

$$y = \Re \int^w i(\Phi^2 + \Psi^2)\, dw,$$

$$z = \Re \int^w 2\,\Phi\Psi\, dw.$$

where $\varPhi$ and $\varPsi$ denote single-valued analytic functions of $w = u + iv$ in $|w| < 1$.

Problem P_4. Solve problem P_3 under the supplementary condition that $\varPhi$, $\varPsi$ do not have any common zero in $|w| < 1$.

Problem P_5. Suppose that the orthogonal projection of the given JORDAN curve $\varGamma^*$ upon the xy-plane is a simply covered JORDAN curve $\varGamma$. Denote by R the JORDAN region bounded by $\varGamma$. Solve problem P_1 under the supplementary condition that the solution admits of a representation $S: z = z(x, y)$, (x, y) in R, where $z(x, y)$ is single-valued and continuous in R and analytic in the interior of R.

Besides the problem of PLATEAU, we shall have to consider *the problem of the least area*, which requires the determination of a continuous surface S bounded by a given JORDAN curve $\varGamma^*$, such that the LEBESGUE area $\mathfrak{A}(S)$ of S is a minimum if compared with the areas of all continuous surfaces bounded by $\varGamma^*$, it being understood that only surfaces of the type of the circular disc are considered[1].

Then comes *the simultaneous problem*, which requires the determination of a common solution of the problem of PLATEAU and of the problem of the least area[2]. According to the different statements of the problem of PLATEAU, we have, strictly speaking, a number of simultaneous problems. As a matter of fact, only the statements P_2 and P_5 of the problem of PLATEAU actually have been used in this connection.

Any one of the preceding problems gives rise to the question as to whether or not the solution is unique. Furthermore, there arises the question as to whether or not some of these problems are equivalent. The existence theorems concerned with the problems listed above will be discussed in subsequent Chapters. The other questions raised in this section will be considered in the present chapter. Sections III.6 to III.16 contain a number of special facts which are then coordinated in the sections III.17 to III.20.

III.6. The following definitions prove useful for the sequel. Let there be given, in a JORDAN region R of the uv-plane, a continuous function $g(u, v)$. Then $g(u, v)$ will be called a *generalized harmonic function* in R if for every interior point (u_0, v_0) of R the following condition is satisfied. There exists a vicinity V_0 of (u_0, v_0) and a topological transformation $u = u_0(\alpha, \beta)$, $v = v_0(\alpha, \beta)$ of V_0 into some region $\overline{V}_0$ of an $\alpha\beta$-plane, such that the function $g[u_0(\alpha, \beta), v_0(\alpha, \beta)]$ of α, β is harmonic in $\overline{V}_0$. Variables α, β with this property will be called *local typical variables* for the point (u_0, v_0).

---

[1] Surfaces of different topological types will be only considered at the end of Chapter VI.

[2] This problem has been called the problem of PLATEAU by LEBESGUE: Intégrale, longueur, aire. Ann. Mat. pura appl. Vol. 7 (1902) pp. 231–359.

Suppose $g(u, v)$ is a generalized harmonic function in R. Suppose that $g(u, v)$ vanishes at an interior point (u_0, v_0) of R. Let (α_0, β_0) be the image of (u_0, v_0) under the transformation $u = u_0(\alpha, \beta)$, $v = v_0(\alpha, \beta)$, where α, β are local typical variables for (u_0, v_0). Suppose that all the partial derivatives of g with respect to α, β vanish at (α_0, β_0) up to and including a certain order $(n-1) \geqq 0$ while at least one of the partial derivatives of order n is different from zero. Then (u_0, v_0) will be called a zero of order n of $g(u, v)$. This notion is independent of the particular choice of the local typical variables α, β.

III.7. If $g_1(u, v)$, $g_2(u, v)$ are generalized harmonic functions in R, then $g_1 + g_2$ generally is not a generalized harmonic function. On the other hand, several important properties of harmonic functions remain valid for generalized harmonic functions. The property expressed by the principle of maximum and minimum clearly remains valid. The following lemma is an immediate consequence of well-known properties of harmonic functions which remain valid for generalized harmonic functions.

Lemma[1]. Let $g(u, v)$ be a generalized harmonic function in a (simply connected) JORDAN region R. Suppose that $g(u, v)$ has a zero (u_0, v_0) of order $n \geqq 1$ in the interior of R. Then $g(u, v)$ vanishes in at least $2n$ distinct points on the boundary of R.

III.8. Consider now a surface $S : \mathfrak{x} = \mathfrak{x}(u, v)$, (u, v) in R, which is a minimal surface (in the sense defined in III.3) bounded by a JORDAN curve Γ^*. If α, β are local typical parameters (see III.3) for an interior point (u_0, v_0) of R, then the components x, y, z of $\mathfrak{x}$ as functions of α, β are harmonic functions. Hence, if a, b, c, d are any four constants, then $ax + by + cz + d$ is also a harmonic function of α, β. Thus if a, b, c, d are any four constants, then the function $ax(u, v) + by(u, v) + cz(u, v) + d$ is a generalized harmonic function in R. The following theorems are immediate consequences of this remark, on account of the facts referred to in III.7.

1. If a convex region K in the xyz-space contains the boundary curve Γ^* of a minimal surface (see III.3) then the whole surface is contained in K[†].

2. The tangent plane, at a regular point (see III.4) of the minimal surface, intersects the boundary curve in at least four distinct points[2].

---

[1] T. RADÓ: Contributions to the theory of minimal surfaces. Acta Litt. Sci. Szeged Vol. 6 (1932) p. 10, where the lemma is stated for $n = 2$.

[†] The reviewer learned about this theorem from L. FEJÉR.

[2] The theorem is also true for surfaces with negative curvature. This fact played an important role in the work of S. BERNSTEIN on partial differential equations of the elliptic type. See for references I. LICHTENSTEIN: Neuere Entwicklung usw. Enzyklopädie der math. Wiss. Vol. 2 (3) pp. 1277—1334.

3. Every plane passing through a branch-point of order n (see III.4) of the minimal surface intersects the boundary curve Γ^* in at least $2(n + 1)$ distinct points[1].

III.9. The last theorem permits us to exclude the possibility of branch-points in certain cases. Suppose there exists, in the xyz-space, a straight line l such that no plane through l intersects the boundary curve Γ^* in more than two distinct points. Then the minimal surface cannot have branch-points, as follows immediately from theorem 3 in III.8. The assumption is, for instance, satisfied if Γ^* has a simply covered star-shaped Jordan curve as its parallel or central projection upon some plane. If the projection has the stronger property of being convex, then a much stronger conclusion can be drawn. Suppose that the parallel projection of Γ^* upon some plane is a simply covered convex curve. Choose a plane perpendicular to the direction of projection for the xy-plane. Then the orthogonal projection of Γ^* upon the xy-plane is again a simply covered convex curve which we shall call Γ. Let

$$S : x = x(u, v), \quad y = y(u, v), \quad z = z(u, v), \quad (u, v) \text{ in } R$$

be the minimal surface under consideration. From theorem 3 in III.8 it follows that S has no branch-points. From theorem 2 in III.8 it follows that S has no tangent plane perpendicular to the xy-plane. From this it follows that S has a simply covered xy-projection in the small. Hence the equations $x = x(u, v)$, $y = y(u, v)$ define a transformation with the following properties.

1. The transformation is one-to-one and continuous in the vicinity of every interior point (u_0, v_0) of R.

2. The boundary of R is carried in a one-to-one and continuous way into the Jordan curve Γ.

On account of the so-called monodromy theorem in topology[2], it follows then that the transformation

$$x = x(u, v), \quad y = y(u, v), \quad (u, v) \text{ in } R$$

carries R in a topological way into the Jordan region bounded by Γ. Hence u, v can be expressed as single-valued continuous functions of x, y in R, and there follows then for the minimal surface S a representation

$$S : z = z(x, y), \quad (x, y) \text{ in or on } \Gamma,$$

where $z(x, y)$ is single-valued and continuous in and on Γ. Since S has no branch-points, S is a minimal surface in the sense of differential geometry. Hence (see II.16) $z(x, y)$ is analytic in the interior of Γ and satisfies there the partial differential equation

$$(1 + q^2) r - 2pqs + (1 + p^2) t = 0. \tag{3.1}$$

---

[1] See T. Radó: The problem of the least area and the problem of Plateau. Math. Z. Vol. 32 (1930) p. 794, where the theorem is stated for $n = 1$.

[2] See, for instance, Kerékjártó: Vorlesungen über Topologie I, p. 175.

Similar conclusions might be obtained if the boundary curve Γ^* is supposed to have a simply covered convex curve as its central projection upon some plane.

III.10. Summing up, we have the following results[1]. Let S be a minimal surface (in the sense of III.3) bounded by a JORDAN curve Γ^*. If Γ^* has a simply covered star-shaped JORDAN curve Γ as its parallel or central projection upon some plane, then S has no branch-points, that is to say S is a minimal surface in the sense of differential geometry. If the projection Γ is convex, then S does not intersect itself even in the large. If the orthogonal projection of Γ^* upon the xy-plane is a simply covered convex curve Γ, then S can be represented in the form $S : z = z(x, y)$, (x, y) in or on Γ, where $z(x, y)$ is single-valued and continuous in and on Γ, and satisfies in the interior of Γ the partial differential equation (3.1).

III.11. We are going to consider now certain *uniqueness theorems*. A first important fact in this connection is the uniqueness theorem for the partial differential equation (3.1). Let there be given, on a JORDAN curve Γ in the xy-plane, a continuous boundary function $\varphi(P)$ of the point P varying on Γ. If then $z_1(x, y)$, $z_2(x, y)$ are solutions of (3.1) which both reduce to $\varphi(P)$ on Γ, then $z_1(x, y) \equiv z_2(x, y)$ in the whole interior of Γ.[2]

Denote then by Γ^* the JORDAN curve, in the xyz-space, determined by the equation $z = \varphi(P)$. The above uniqueness theorem asserts that Γ^* cannot bound more than one minimal surface which has a simply covered xy-projection. This statement is rather unsatisfactory; indeed, we shall see (III.17) that the boundary curve of a minimal surface might very well have a simply covered xy-projection, while the minimal surface itself does not have this property. On the other hand, if the xy-projection of the boundary curve Γ^* is convex, then every minimal surface bounded by Γ^* has also a simply-covered xy-projection, on account of III.10. A similar argument holds in case Γ^* is known to have a simply-covered convex curve as its central projection. In this way results the following uniqueness theorem[3].

If a JORDAN curve Γ^* has a simply covered convex curve as its parallel or central projection upon some plane, then Γ^* cannot bound more than

---

[1] See T. RADÓ: The problem of the least area and the problem of PLATEAU. Math. Z. Vol. 32 (1930) pp. 763—796. — T. RADÓ: Contributions to the theory of minimal surfaces. Acta Litt. Sci. Szeged Vol. 6 (1932) pp. 1—20.

[2] See, also for references, the beautiful treatment of this theorem and of related subjects by E. HOPF: Elementare Bemerkungen über die Lösungen partieller Differentialgleichungen zweiter Ordnung vom elliptischen Typus. S.-B. preuß. Akad. Wiss. 1927 pp. 147—152. See also A. HAAR: Über reguläre Variationsprobleme. Acta Litt. Sci. Szeged Vol. 3 (1927) pp. 224—234.

[3] T. RADÓ: Contributions to the theory of minimal surfaces. Acta Litt. Sci. Szeged Vol. 6 (1932) pp. 1—20.

one minimal surface (this term being used in the general sense defined in III.3).

III.12. The assumptions of the preceding uniqueness theorem obviously are satisfied if Γ^* is a skew quadrilateral. For this case, the uniqueness theorem has been stated, without proof, by H. A. SCHWARZ. For the purpose of an application to be made later on we mention the following consequence of the uniqueness theorem. Suppose a JORDAN curve Γ^* is invariable under reflection upon a certain plane p. Every minimal surface, bounded by Γ^*, is carried by the reflection into a minimal surface bounded by Γ^*. Hence, if it is known that Γ^* bounds just one minimal surface S, then it follows that S is also invariable under the reflection. Repeated application of this remark to the case when Γ^* consists of the edges AB, BC, CD, DA of a regular tetrahedron with vertices A, B, C, D, leads to the result that the minimal surface bounded by Γ^* passes through the center of the tetrahedron (this fact has been verified by SCHWARZ by using the explicit formulas for the surface).

III.13. Let us consider now the relation between the problem of the least area and the simultaneous problem (see III.5). If a solution S of the problem of the least area satisfies the assumptions made in the classical Calculus of Variations, then S is a minimal surface (see II.5). Without those assumptions this conclusion in general does not hold. The following example is a slight modification of one given by LEBESGUE in his Thesis[1].

Let the given JORDAN curve Γ^* coincide with the circle

$$\Gamma^* : x^2 + y^2 = 1, \quad z = 0.$$

Then the area $\mathfrak{A}(S)$ of any continuous surface S (of the type of the circular disc) bounded by Γ^* is $\geq \pi$. Consider then the surface

$$S: \begin{cases} x = 2^n (r - \tfrac{1}{2})^n \cos\varphi, & y = 2^n (r - \tfrac{1}{2})^n \sin\varphi, & z = 0 \text{ for } \tfrac{1}{2} \leq r \leq 1, \\ x = 0, & y = 0, & z = (1 - 4r^2)^n \text{ for } 0 \leq r \leq \tfrac{1}{2}, \end{cases}$$

where r, φ are polar coordinates in the uv-plane, and n is a given positive integer. Using the relations $u = r\cos\varphi$, $v = r\sin\varphi$, we have the equations of S appearing in the form

$$S: x = x(u, v), \quad y = y(u, v), \quad z = z(u, v), \quad u^2 + v^2 \leq 1,$$

where $x(u, v)$, $y(u, v)$, $z(u, v)$ easily are seen to have continuous partial derivatives up to and including the order $n - 1$. S consists of the simply covered disc $x^2 + y^2 \leq 1$, $z = 0$ and of the spine $x = 0$, $y = 0$, $0 \leq z \leq 1$. The area $\mathfrak{A}(S)$ is found to be equal to π, by computing the integral $\int\int (EG - F^2)^{\frac{1}{2}}$. Hence the area of S is a minimum. Still, S is not a minimal surface, not even in the general sense defined in III.3. Indeed, a minimal surface is comprised in every convex region

[1] Intégrale, longueur, aire. Ann. Mat. pura appl. Vol. 7 (1902) pp. 231—359.

which contains its boundary curve (see III.8), and S obviously does not satisfy this condition.

Instead of using one spine as above, we can disfigure any given surface S by putting on it any finite number of spines, without changing its area. We shall see (Chapter VI) that the problem of the least area is solvable for every JORDAN curve. From these two facts it follows that *the problem of the least area has infinitely many solutions for every* JORDAN *curve* Γ^*, *and that a surface S which solves the problem is not in general a minimal surface.*

III.14. Thus the solution of the problem of the least area does not imply the solution of the problem of PLATEAU. Neither does the solution of the problem of PLATEAU imply the solution of the problem of the least area; in other words, *the area of a minimal surface, bounded by a given* JORDAN *curve* Γ^*, *is not necessarily a minimum.* This fact has been recognized at the earliest stage of the theory. For the case of doubly connected minimal surfaces bounded by two given curves, the catenoids offer simple examples for the lack of the minimizing property[1]. For the case of minimal surfaces, of the type of the circular disc, bounded by a given curve, H. A. SCHWARZ obtained very general examples in the following way[2].

Consider, in the $u + iv = w$ plane, a JORDAN region R bounded by an analytic JORDAN curve. Denote by $\mu(w)$ a function which is analytic and different from zero in R. Then the equations

$$\left. \begin{aligned} x &= \Re \int^{w} (1 - w^2)\,\mu(w)\,dw, \\[4pt] y &= \Re \int^{w} i(1 + w^2)\,\mu(w)\,dw, \\[4pt] z &= \Re \int^{w} 2\,w\,\mu(w)\,dw, \end{aligned} \right\} \tag{3.2}$$

where w varies in R, define a regular minimal surface[3]. The area of this surface certainly is not a minimum if the inequality (2.4) in II.7 is not satisfied. The left-hand side of that inequality reduces in the present case to

$$\iint\limits_{R} \left[\lambda_u^2 + \lambda_v^2 - \frac{8\,\lambda^2}{(1 + u^2 + v^2)^2} \right] du\, dv. \tag{3.3}$$

Hence the area of the minimal surface certainly is not a minimum if this integral can be made negative by substituting a function $\lambda(u, v)$ which has continuous partial derivatives of the first order in R and which vanishes on the boundary of R.

---

[1] See, also for references, the beautiful Chapter IV in the little book of G. A. BLISS: Calculus of Variations (No. 1 of the Carus Mathematical Monographs).

[2] Gesammelte Mathematische Abhandlungen Vol. 1 pp. 151—167 and 223—269.

[3] The formulas (3.2) are obtained by choosing $\lambda(w) = w$ in (2.28).

This criterion, curiously enough, does not depend upon the function $\mu(w)$ which determines the minimal surface; the criterion is concerned solely with the region R. SCHWARZ based the discussion of this situation on the study of the characteristic values of a certain partial differential equation[1]. One of the results he obtained states that if the region R contains the unit circle $u^2 + v^2 \leq 1$ in its interior, then the integral (3.3) can be made negative. On account of the geometrical meaning[2] of the variable w in the formulas (3.2), this assumption concerning R means that the spherical image of the minimal surface defined by (3.2) completely covers half the unit sphere.

This result of SCHWARZ can be obtained also in the following elementary way[3]. Let r be a positive parameter and define a function $\lambda(u, v; r)$ by

$$\lambda(u, v; r) = \frac{u^2 + v^2 - r^2}{u^2 + v^2 + r^2} \quad \text{for} \quad 0 \leq u^2 + v^2 \leq r^2.$$

Put

$$J(r) = \iint\limits_{u^2 + v^2 < r^2} \left[\lambda_u^2 + \lambda_v^2 - \frac{8\lambda^2}{(1 + u^2 + v^2)^2} \right] du\, dv.$$

Partial integration gives

$$J(r) = -\iint\limits_{u^2 + v^2 < r^2} \lambda \left[\Delta\lambda + \frac{8\lambda}{(1 + u^2 + v^2)^2} \right] du\, dv.$$

Actual computation shows that the function $\lambda(u, v; 1)$ satisfies the equation

$$\Delta\lambda + \frac{8\lambda}{(1 + u^2 + v^2)^2} = 0.$$

Hence $J(1) = 0$. We want now to determine $J'(1)$. Changing to new variables

$$\alpha = \frac{u}{r}, \quad \beta = \frac{v}{r},$$

we transform $J(r)$ into a new integral taken over the fixed region $\alpha^2 + \beta^2 < 1$. $J'(1)$ can then be computed by differentiating under the integral sign. The result is

$$J'(1) = 16 \iint\limits_{\alpha^2 + \beta^2 < 1} \frac{(\alpha^2 + \beta^2 - 1)^3}{(\alpha^2 + \beta^2 + 1)^5} d\alpha\, d\beta.$$

Since the integrand obviously is negative, it follows that $J'(1) < 0$. From $J(1) = 0$, $J'(1) < 0$ it follows that we have a constant $\sigma > 1$, such that $J(r) < 0$ for $1 < r < \sigma$.

Suppose then that a JORDAN region R contains $u^2 + v^2 \leq 1$ in its interior. Then r can be determined in such a way that $1 < r < \sigma$ and

[1] Gesammelte Mathematische Abhandlungen Vol. 1 pp. 241—269.

[2] w is the stereographic projection of the spherical image of the surface. See DARBOUX: Théorie générale des surfaces Vol. 1 pp. 347—348.

[3] T. RADÓ: Contributions to the theory of minimal surfaces. Acta Litt. Sci. Szeged Vol. 6 (1932) pp. 1—20.

that the circle $u^2 + v^2 \leqq r^2$ is interior to R. Define a function $\lambda(u, v)$ by the formulas

$$\lambda(u, v) = \begin{cases} \lambda(u, v; r) & \text{in} \quad 0 \leqq u^2 + v^2 \leqq r^2, \\ 0 & \text{for} \quad u^2 + v^2 > r^2. \end{cases}$$

Then $\lambda(u, v)$ vanishes on the boundary of R and makes the integral (3.3) negative. The first partial derivatives of $\lambda(u, v)$ are discontinuous on $u^2 + v^2 = r^2$; this edge however can be rounded off by a familiar process.

It is thus proved that if R contains $u^2 + v^2 \leqq 1$ in its interior, then the formulas (3.2) define, for every choice of the analytic function $\mu(w)$, a minimal surface the area of which is not a minimum.

III.15. In order to obtain a clear-cut example, the function $\mu(w)$ in (3.2) should be chosen in such a way that the resulting minimal surface be bounded by a JORDAN curve. It can easily be shown that for $\mu(w) \equiv \tfrac{1}{3}$ this condition is satisfied, provided R is a circle $u^2 + v^2 \leqq r^2$ with $r < \sqrt{3}$. Then the following explicit example is obtained[1]. *The equations*

$$\left. \begin{aligned} x &= u + uv^2 - \tfrac{1}{3}u^3, \\ y &= -v - u^2v + \tfrac{1}{3}v^3, \\ z &= u^2 - v^2, \end{aligned} \right\} \quad u^2 + v^2 \leqq r^2, \quad 1 < r < \sqrt{3}, \qquad (3.4)$$

define a minimal surface S bounded by a JORDAN *curve, such that the area of S is not a minimum*[2].

III.16. The area of the minimal surface S given by (3.4) clearly is finite. On account of a general existence theorem to be considered in Chapter VI, it follows that the boundary curve of S bounds a minimal surface S^* the area of which is a minimum. Then S^* is not identical to S, because the area of S is not a minimum. Hence the boundary curve of S bounds at least two distinct minimal surfaces. The equations of the boundary curve, in terms of polar coordinates r, Θ, where $u = r\cos\Theta$, $v = r\sin\Theta$, are obtained in the form

$$\left. \begin{aligned} x &= r \cos\Theta - \tfrac{1}{3}r^3 \cos 3\Theta, \\ y &= -r \sin\Theta - \tfrac{1}{3}r^3 \sin 3\Theta, \\ z &= r^2 \cos 2\Theta. \end{aligned} \right\} \qquad 0 \leqq \Theta < 2\pi. \qquad (3.5)$$

Thus we have the following example[1]: *if* $1 < r < \sqrt{3}$, *the equations* (3.5) *define a* JORDAN *curve which bounds at least two distinct minimal surfaces of the type of the circular disc.*

---

[1] T. RADÓ: Contributions to the theory of minimal surfaces. Acta Litt. Sci. Szeged Vol. 6 (1932) pp. 1—20.

[2] The formulas (3.4) define the so-called minimal surface of ENNEPER. See DARBOUX: Théorie générale des surfaces Vol. 1 pp. 372—376. Examples of a less elementary character have been given by H. A. SCHWARZ: Gesammelte Mathematische Abhandlungen Vol. 1 pp. 151—167 and 223—269.

The catenoids give explicit examples of distinct minimal surfaces with identical boundaries[1]. It seems that no explicit example has yet been given for two minimal surfaces of the type of the circular disc and bounded by the same JORDAN curve. In the above example, one of the two minimal surfaces is given explicitly by the formulas (3.4), while the other is only known to *exist*[2]. At any rate, *the solution of the problem of* PLATEAU *in general is not unique*; on the other hand, the solution is unique for certain special classes of curves (see III.11). It would be desirable to obtain more comprehensive information concerning those properties of the given boundary curve which determine the number of the solutions.

III.17. We shall now discuss briefly the statements of the problems P_1 to P_5, listed in III.5. While problem P_1 simply calls for a minimal surface S (of the type of the circle) bounded by a given JORDAN curve Γ^*, problems P_2 to P_5 require that S admits of a representation of a prescribed type. In the case of problem P_5, this implies a restriction on Γ^* also, namely the restriction that Γ^* admits of a simply covered xy-projection.

In the course of general investigations on partial differential equations of the elliptic type, S. BERNSTEIN[3] observed that a large class of problems, including as a special case our problem P_5, in general is not solvable. The following example[4] shows the geometrical reasons for this fact for problem P_5. Take a regular tetrahedron, with vertices A, B, C, D, and let the xy-plane pass through A, B, C. Denote by Γ^* the quadrilateral AB, BC, CD, DA. Then Γ^* has a simply covered xy-projection. Yet, problem P_5 does not have a solution for Γ^*. There certainly exists a minimal surface S bounded by Γ^*, as has been proved already by SCHWARZ. We also know (III.11) that S is the only minimal surface (of the type of the circle) bounded by Γ^*. S passes through the center O of the tetrahedron (III.12), and thus S contains two points, namely O and D, with the same xy-projection. Hence *there exists a unique minimal surface bounded by Γ^*, and this minimal surface does not satisfy the additional condition of having a simply covered xy-projection*[5].

---

[1] See G. A. BLISS: Calculus of Variations (No. 1 of the Carus Mathematical Monographs).

[2] The same remark applies to an example due to N. WIENER. See J. DOUGLAS: Solution of the problem of PLATEAU. Trans. Amer. Math. Soc. Vol. 33 (1931) p. 269.

[3] S. BERNSTEIN: Sur les équations du Calcul des Variations. Ann. École norm. Vol. 29 (1912) pp. 431—485. See in particular pp. 484—485.

[4] T. RADÓ: Contributions to the theory of minimal surfaces. Acta Litt. Sci. Szeged Vol. 6 (1932) pp. 1—20.

[5] The *uniqueness* is absolutely essential for the conclusion that problem P_5 is not solvable. For this reason, several examples presented in the literature are incomplete.

This example shows very clearly that the trouble comes from the fact that problem P_5 is a *non-parametric problem,* inasmuch as it calls for a surface represented by an equation of the form $z = z(x, y)$, instead of asking for a surface in the general parametric form

$$x = x(u, v), \quad y = y(u, v), \quad z = z(u, v).$$

III.18. The problems P_1 to P_4 (see III.5) are statements of *the problem of* PLATEAU *in the parametric form.* Problems P_2, P_3, P_4 call, roughly speaking, for a minimal surface in terms of *isothermic parameters in the large.* The character of this condition might be well illustrated in the special case when Γ^* is a plane curve.

Suppose Γ^* is in the xy-plane, and consider, for instance, a solution

$$S : x = x(u, v), \quad y = y(u, v), \quad z = z(u, v), \quad u^2 + v^2 \leqq 1$$

of problem P_2. Then $z(u, v) = 0$ on $u^2 + v^2 = 1$. As $z(u, v)$ is harmonic, it follows that $z(u, v) \equiv 0$. The conditions $E = G$, $F = 0$ reduce therefore to

$$x_u^2 + y_u^2 = x_v^2 + y_v^2, \quad x_u x_v + y_u y_v = 0.$$

This implies that $x(u, v)$ and $y(u, v)$ are conjugate harmonic functions, that is to say real and imaginary parts of an analytic function $f(w)$ of $w = u + iv$. The statement of problem P_2 requires that

1. $f(w)$ is analytic for $|w| < 1$, and
2. $f(w)$ is continuous for $|w| \leqq 1$, and
3. the equation $\zeta = f(w)$ carries $|w| = 1$ in a topological way into the JORDAN curve Γ^* of the $\zeta = x + iy$ plane.

On account of a classical theorem of DARBOUX (see V.19), this situation implies that the equation $\zeta = f(w)$ carries $|w| < 1$ in a one-to-one and conformal way into the interior of Γ^*.

Hence problem P_2 requires the mapping of the JORDAN region, bounded by Γ^*, in a one-to-one and continuous, and in the interior conformal, way upon the unit circle $|w| \leqq 1$. The same holds obviously for problems P_3 and P_4. While problems P_2, P_3, P_4 reduce, in the case of a curve situated in the xy-plane, to one of the fundamental problems in conformal mapping, problem P_1 simply is trivial in this case. Indeed, if R denotes the JORDAN region bounded by Γ^*, then the surface

$$S : x = x, \ y = y, \ z = 0, \ (x, y) \text{ in } R$$

clearly solves problem P_1.

It might be inferred from these remarks that problem P_1 is *easier* to solve than problem P_2. Curiously enough, nobody ever has tried to take advantage of this possibility. The problem of PLATEAU, in the parametric form, has always been asked in one of the statements P_2, P_3, P_4 [1] which also require the mapping of the minimal surface conformally upon the unit circle in the large.

---

[1] Or in the even more severe form based on the formulas (3.2).

The question naturally arises whether or not every solution of problem P_1 is also a solution of problem P_2. This amounts to the question whether or not a minimal surface (in the general sense of III.3) bounded by a JORDAN curve Γ^* always admits of a representation as specified by the statement of problem P_2. It seems that the methods developed for dealing with the OSGOOD-CARATHÉODORY theorem[1] will permit, after proper adjustments, the answering of this question in the affirmative[2].

III.19. Consider now a solution

$$S : x = x(u, v), \quad y = y(u, v), \quad z = z(u, v), \quad u^2 + v^2 \leqq 1$$

of problem P_3 or P_4 (see III.5). We have then the equations

$$\left. \begin{array}{l} x_u - i x_v = \Phi^2 - \Psi^2, \\ y_u - i y_v = i(\Phi^2 + \Psi^2), \\ z_u - i z_v = 2\,\Phi\,\Psi. \end{array} \right\} \tag{3.6}$$

The condition that $EG - F^2 = 0$ for an interior point (u_0, v_0) is then readily found to be equivalent to $\Phi = 0$, $\Psi = 0$ for $w = w_0 = u_0 + i v_0$. Since problem P_4 requires that Φ and Ψ have no common zeros, we see that *in problem P_4 the solution is not permitted to have branch-points.* It easily is seen that if a solution of problem P_3 does not have branch-points, then it admits of a representation as required in problem P_4. In other words, if we add the condition $EG - F^2 > 0$ for $u^2 + v^2 < 1$ in the statement of problem P_2, then we obtain problem P_4, that is to say the classical statement of the problem of PLATEAU in the parametric form[3].

The question arises whether or not problem P_4 is always possible. The impossibility, in general, of the problem would be demonstrated by exhibiting a single JORDAN curve Γ^* for which it could be proved that *every* minimal surface (of the type of the circular disc) bounded by Γ^* necessarily has branch-points. Such a curve has not yet been exhibited; the surmisal that any knotted JORDAN curve would serve the purpose can readily be refuted by examples of knotted JORDAN curves which do bound minimal surfaces (of the type of the circular disc) free of branch-points. More explicitly: there exist knotted JORDAN curves for which the classical problem P_4 is possible, and no JORDAN curve is known at present for which problem P_4 is impossible[4].

---

[1] We mean the following theorem: if the interior of a JORDAN curve is mapped, in a one-to-one and conformal way, upon the interior of the unit circle, then the map remains continuous and one-to-one on the boundaries. See for instance CARATHÉODORY: Conformal representation (Cambridge Tracts 28).

[2] This program has been carried out in a joint paper by E. F. BECKENBACH and T. RADÓ: Subharmonic functions and minimal surfaces. To appear in Trans. Amer. Math. Soc.

[3] See H. A. SCHWARZ: Gesammelte Mathematische Abhandlungen Vol. I.

[4] Cf. VI.35.

Combining the existence theorem in Chapter V with the theorems in III.10, we obtain existence theorems for the classical problem P_4 which seem to be the most general known at present. For instance: if a JORDAN curve Γ^* has a simply covered star-shaped curve as its parallel or central projection upon some plane, then problem P_4 is solvable for Γ^*.

III.20. While problem P_4 excludes branch-points altogether, and while problem P_2 does not imply any restriction as to branch-points, the geometrical interpretation of problem P_3 is less clear-cut.

Differentiating again the equations (3.6), we obtain

$$\left.\begin{aligned}
x_{uu} - i x_{uv} &= 2(\Phi \Phi' - \Psi \Psi'), \\
y_{uu} - i y_{uv} &= 2i(\Phi \Phi' + \Psi \Psi'), \\
z_{uu} - i z_{uv} &= 2(\Phi \Psi' + \Phi' \Psi).
\end{aligned}\right\} \qquad (3.7)$$

From (3.6) and (3.7) it follows readily: if, for a solution of problem P_3, we have $EG - F^2 = 0$ at an interior point $u_0 + i v_0 = w_0$, then the partial derivatives of the first and second order of $x(u, v)$, $y(u, v)$, $z(u, v)$ all vanish at that point. In other words: *the branch-points of a solution of problem P_3 are at least of order 2* (see III.4 for the definition). On the other hand, examples show that problem P_2 might have solutions with branch-points of order one.

Thus it follows that *the problems P_2 and P_3 are not equivalent.* Problem P_3 will be seen to be solvable for every not knotted JORDAN curve (Chapter V). Again, it is not known at the present time whether there do or do not exist curves for which problem P_3 is impossible.

III.21. We have seen (III.18) that if the given JORDAN curve Γ^* is situated in the xy-plane, then the problem of PLATEAU in the parametric form (in any one of the statements P_2, P_3, P_4 of III.5) reduces to the problem of mapping the JORDAN region bounded by Γ^* in a one-to-one and continuous, and in the interior conformal, way upon the unit circle $u^2 + v^2 \leq 1$. This situation played an important role in the theory. The most direct illustration is given by the case when Γ^* is a *polygon*. One of the earliest ideas for dealing with the problem of the conformal mapping of the unit circle upon the region bounded by a plane polygon was based on the principle of symmetry and led to the so-called formulas of SCHWARZ and CHRISTOFFEL. The method used by SCHWARZ in his classical investigations on the problem of PLATEAU is clearly a generalization to the space of the plane method[1]. In a general way, the reader will find, in the Chapters IV, V, VI dealing with the existence theorems, many instances where the theory of the conformal mapping of plane

---

[1] Compare, for instance, the following two papers of H. A. SCHWARZ: Über einige Abbildungsaufgaben. Gesammelte Mathematische Abhandlungen Vol. 2 pp. 65—83 and Bestimmung einer speziellen Minimalfläche. Gesammelte Mathematische Abhandlungen Vol. 1 pp. 6—125. See also DARBOUX: Théorie générale des surfaces Vol. 1 pp. 490—601.

regions clearly served as a model for the development of the theory of the problem of PLATEAU.

An important remark, due to J. DOUGLAS, should be mentioned here. Since the problem of OSGOOD-CARATHÉODORY (conformal mapping of a plane JORDAN region upon the circle) is included in problem P_2, it follows that if we have a method of solution of problem P_2 which does not make use of the solution of that problem, then we have a simultaneous solution of the problem of OSGOOD-CARATHÉODORY and of the problem of PLATEAU. J. DOUGLAS emphasizes the fact that this is the case with his own method[1].

III.22. Instead of considering only existence theorems, the relation between conformal mapping of plane regions, that is to say analytic functions of a complex variable, on the one side and minimal surfaces on the other side might be discussed on account of its own intrinsic interest. The theory of minimal surfaces appears then as a generalization of the theory of analytic functions of a complex variable. While the Chapters IV, V, VI will review numerous facts and methods which might be interpreted from this point of view, it might be useful to present to the reader some specific illustrations.

III.23. If $f(w) = x(u, v) + iy(u, v)$ is an analytic function of w for $|w| < 1$, then $x_u = y_v$, $x_v = -y_u$ (CAUCHY-RIEMANN equations). From these equations it follows that

$$x_u^2 + y_u^2 = x_c^2 + y_c^2, \qquad x_u x_v + y_u y_v = 0. \tag{3.8}$$

Conversely, it follows from (3.8) that either y is conjugate harmonic to x, or x is conjugate harmonic to y. Let us call two harmonic functions related by (3.8) *a couple of conjugate harmonic functions.*

On the other hand, the theorem of WEIERSTRASS (see II.17) leads to consider triples of harmonic functions related by the equations

$$x_u^2 + y_u^2 + z_u^2 = x_c^2 + y_c^2 + z_c^2, \qquad x_u x_v + y_u y_v + z_u z_v = 0. \tag{3.9}$$

We shall say that x, y, z form *a triple of conjugate harmonic functions.* We shall review now a few facts which develop further this analogy.

III.24. Suppose $f(w)$ is analytic in $|w| < 1$ and even on $|w| = 1$, for the sake of simplicity. Put $f'(w) = g(w)$. Then the area of the image of $|w| \leqq 1$ by $f(w)$ is given by

$$\mathfrak{A} = \int_0^1 \int_0^{2\pi} |g(re^{i\Theta})|^2 r \, dr \, d\Theta, \tag{3.10}$$

while the length L of the image of $|w| = 1$ is

$$L = \int_0^{2\pi} |g(e^{i\Theta})| \, d\Theta. \tag{3.11}$$

---

[1] See J. DOUGLAS· Solution of the problem of PLATEAU. Trans. Amer. Math. Soc. Vol. 33 (1931) pp. 263—321.

If $f(w)$ would give a simply covered image of $|w| \leq 1$, then we could assert, on account of the isoperimetric inequality, that

$$\mathfrak{A} \leq \frac{1}{4\pi} L^2,$$

or, on account of (3.10) and (3.11), that

$$\int_0^1 \int_0^{2\pi} |g(re^{i\Theta})|^2 r\, dr\, d\Theta \leq \frac{1}{4\pi} \left(\int_0^{2\pi} |g(e^{i\Theta})|\, d\Theta \right)^2. \tag{3.12}$$

CARLEMAN[1] proved that (3.12) holds regardless if $f(w)$ gives a simply covered image. From this he inferred that between the area $\mathfrak{A}$ of a minimal surface and the length L of its boundary curve the isoperimetric inequality also holds. To prove this[2], suppose, to simplify the discussion, that the minimal surface is given by

$$S: x = x(u, v), \quad y = y(u, v), \quad z = z(u, v), \quad u^2 + v^2 \leq 1, \tag{3.13}$$

where x, y, z form a triple of conjugate harmonic functions (see III.23) which remain analytic even on $u^2 + v^2 = 1$. Then x, y, z are the real parts of analytic functions:

$$x = \Re f_1(w), \quad y = \Re f_2(w), \quad z = \Re f_3(w),$$

and if we put $f_1' = g_1, f_2' = g_2, f_3' = g_3$ and use the equation $g_1^2 + g_2^2 + g_3^2 = 0$ (see II.18), then the isoperimetric inequality is expressed by

$$\frac{1}{2} \sum_{k=1}^{3} \int_0^1 \int_0^{2\pi} |g_k(re^{i\Theta})|^2 r\, dr\, d\Theta \leq \frac{1}{4\pi} \left[\int_0^{2\pi} \left(\frac{1}{2} \sum_{k=1}^{3} |g_k(e^{i\Theta})|^2 \right)^{\frac{1}{2}} d\Theta \right]^2. \tag{3.14}$$

To prove (3.14), observe that on account of the inequality of MINKOWSKI[3] we have

$$\frac{1}{2} \sum_{k=1}^{3} \left(\int_0^{2\pi} |g_k(e^{i\Theta})|\, d\Theta \right)^2 \leq \left[\int_0^{2\pi} \left(\frac{1}{2} \sum_{k=1}^{3} |g_k(e^{i\Theta})|^2 \right)^{\frac{1}{2}} d\Theta \right]^2. \tag{3.15}$$

Hence (3.14) follows immediately from (3.15) and (3.12). Some more discussion shows that the sign of equality in (3.14) holds if and only if the minimal surface reduces to a simply covered circular disc.

If we suppose that the minimal surface has a minimum area, then the theorem of CARLEMAN is almost trivial, as it has been observed by BLASCHKE[4]. Indeed, consider a cone consisting of the straight segments which connect a fixed point of the boundary curve with a variable point on that curve. Since cones are developable, the isoperimetric inequality holds for this cone, and hence all the more for the minimal surface, since the area of this latter is by assumption not greater than the area

---

[1] Zur Theorie der Minimalflächen. Math. Z. Vol. 9 (1921) pp. 154—160.

[2] We follow the simple proof given by E. F. BECKENBACH: The area and boundary of minimal surfaces. Ann. of Math. Vol. 33 (1932) pp. 658—664.

[3] See for instance PÓLYA-SZEGÖ: Aufgaben und Lehrsätze Vol. 2 p. 14.

[4] T. CARLEMAN: Zur Theorie der Minimalflächen. Math. Z. Vol. 9 (1921) p. 160.

of the cone, while the boundary curve is the same for either surface. This remark of BLASCHKE shows that the point to the theorem of CARLE-MAN is that the theorem is true even if the area of the minimal surface is not a minimum (cf. III.14).

III.25. Further inequalities between the area $\mathfrak{A}$ of a minimal surface and the length L of its boundary curve have been obtained by BECKEN-BACH[1]. Suppose that the minimal surface is again given by the equations (3.13). Suppose also that at the origin we have $E = 1$ (that is to say, the linear ratio of magnification at the origin is unity). Then

$$\mathfrak{A} \geqq \pi, \qquad L \geqq 2\pi.$$

That is to say, $\mathfrak{A}$ is at least equal to the area and L is at least equal to the perimeter of the unit circle. The sign of equality holds if and only if the minimal surface is a simply covered circular disc. These and similar theorems are proved by BECKENBACH by using FOURIER expansions.

III.26. A minimal surface S being again given by (3.13) where x, y, z are supposed to form a triple of conjugate harmonic functions, we shall call $(x^2 + y^2 + z^2)^{\frac{1}{2}}$ the *norm* of S and we shall write $(x^2 + y^2 + z^2)^{\frac{1}{2}} = |S|$. Then $|S|$ is the generalization of the absolute value of an analytic function $f(w)$ of w. If $f(w)$ is analytic, then $\log|f(w)|$ is a harmonic function of u, v, and this accounts for many important facts in the theory of functions of a complex variable. While $\log|S|$ is not harmonic, it can be shown to be *subharmonic*[2]. A function $g(u, v)$ is subharmonic in a domain D if for every point (u_0, v_0) of D the inequality

$$g(u_0, v_0) \leqq \frac{1}{2\pi} \int_0^{2\pi} g(u_0 + \varrho\cos\Theta, v_0 + \varrho\sin\Theta)\, d\Theta$$

is satisfied for sufficiently small values of ϱ[3]. If g has continuous partial derivatives of the second order, then this condition is equivalent to $\Delta g = g_{uu} + g_{vv} \geqq 0$ and $\log|S|$ is easily shown to satisfy this latter condition. It is sufficient to consider the situation in domains D where $|S| > 0$. Put $\log|S| = g$. Direct computation gives

$$\Delta g = \frac{(\mathfrak{x}_u^2 + \mathfrak{x}_v^2)\,\mathfrak{x}^2 - 2\,((\mathfrak{x}\,\mathfrak{x}_u)^2 + (\mathfrak{x}\,\mathfrak{x}_v)^2)}{|S|^4}, \tag{3.16}$$

[1] See E. F. BECKENBACH: The area and boundary of minimal surfaces. Ann. of Math. Vol. 33 (1932) pp. 658—664. — The theorems in III.24 and III.25 also hold for surfaces of negative curvature, given in isothermic representation. See E. F. BECKENBACH and T. RADÓ: Subharmonic functions and surfaces of negative curvature. To appear in Trans. Amer. Math. Soc.

[2] III.26 to III.29 are taken from E. F. BECKENBACH and T. RADÓ: Subharmonic functions and minimal surfaces. To appear in Trans. Amer. Math. Soc.

[3] See F. RIESZ: Sur les fonctions subharmoniques etc. Acta Math. Vol. 48 (1926) pp. 329—343.

where $\mathfrak{x} = \mathfrak{x}(u, v)$ is the vector equation of S. Since the components of $\mathfrak{x}$ form a triple of conjugate harmonic functions, we have

$$\mathfrak{x}_u^2 = \mathfrak{x}_v^2, \quad \mathfrak{x}_u \mathfrak{x}_v = 0.$$

At those points where $\mathfrak{x}_u^2 = \mathfrak{x}_v^2 = 0$, we have $\Delta g = 0$. At those points where $\mathfrak{x}_u^2 = \mathfrak{x}_v^2 > 0$, $\mathfrak{x}_u$ and $\mathfrak{x}_v$ are different from zero and are perpendicular to each other. Using a unit vector ξ perpendicular to $\mathfrak{x}_u$ and $\mathfrak{x}_v$, we can write

$$\mathfrak{x} = a\mathfrak{x}_u + b\mathfrak{x}_v + c\xi,$$

where a, b, c are scalars. It follows that

$$\mathfrak{x}\mathfrak{x}_u = a\lambda, \quad \mathfrak{x}\mathfrak{x}_v = b\lambda, \quad \mathfrak{x}^2 = \lambda(a^2 + b^2) + c^2, \quad \text{where} \quad \lambda = \mathfrak{x}_u^2 = \mathfrak{x}_v^2.$$

Substituting in (3.16), we get

$$\Delta g = \frac{2\lambda c^2}{|S|^4} \geqq 0.$$

Thus $\Delta g \geqq 0$ always.

III.27. The fact that $\log|S|$ is subharmonic makes it possible to extend a great number of theorems on analytic functions to minimal surfaces, namely those theorems which depend essentially on the fact that the product of analytic functions is again an analytic function. While there is no direct analogy for minimal surfaces, the subharmonic character of $\log|S|$ permits to extend the proofs. We just mention two examples.

Let the minimal surface S be given by the equations (3.13) where x, y, z form a triple of conjugate harmonic functions. Suppose that

    1. $x(0, 0) = 0, \quad y(0, 0) = 0, \quad z(0, 0) = 0,$

    2. $(x^2 + y^2 + z^2)^{\frac{1}{2}} = |S| \leqq 1 \quad \text{in} \quad u^2 + v^2 < 1.$

Then

$$(x(u, v)^2 + y(u, v)^2 + z(u, v)^2)^{\frac{1}{2}} \leqq (u^2 + v^2)^{\frac{1}{2}} \quad \text{in} \quad u^2 + v^2 < 1,$$

and the sign of equality holds if and only if the surface is a simply covered circular disc. This generalizes the lemma of SCHWARZ.

III.28. Suppose this time that the minimal surface is given by

$$S : x = x(u, v), \quad y = y(u, v), \quad z = z(u, v), \quad 0 < \operatorname{arctg} \frac{v}{u} < \alpha,$$

where x, y, z form a triple of conjugate harmonic functions. Suppose these functions remain continuous for $v = 0$, $u > 0$, and suppose that $x(u, 0)$, $y(u, 0)$, $z(u, 0)$ approach definite finite limits x_0^+, y_0^+, z_0^+ for $u \to +0$. Then, in every angle $0 \leqq \operatorname{arctg} \frac{v}{u} \leqq \alpha - \varepsilon$, $\varepsilon > 0$, the functions $x(u, v)$, $y(u, v)$, $z(u, v)$ approach the limits x_0^+, y_0^+, z_0^+ for $(u, v) \to (0, 0)$. This generalizes a well-known theorem of LINDELÖF. The proof follows by an extension of the so-called *multiplication method*[1].

---

[1] See for instance PÓLYA-SZEGÖ: Aufgaben und Lehrsätze Vol. 1 p. 138 problem 277.

III.29. The theorem of III.26 can be completed as follows. *Three functions* $x(u, v)$, $y(u, v)$, $z(u, v)$, *continuous in a domain D, form there a triple of conjugate harmonic functions if and only if* $\log[(x + a)^2 + (y + b)^2 + (z + c)^2]$ *is subharmonic for every choice of the constants a, b, c.*

Chapter IV.

The non-parametric problem.

IV.1. The problem of PLATEAU in the non-parametric form asks for a minimal surface bounded by a given curve and represented as a whole by an equation of the form $z = z(x, y)$. The exact statement of the problem is as follows. Given, in the xy-plane, a JORDAN curve Γ, and a continuous function $\varphi(P)$ of the point P varying on Γ. Determine a function $z(x, y)$ which is continuous in and on Γ, which has continuous derivatives of the first and second orders inside of Γ, which reduces to $\varphi(P)$ on Γ, and which satisfies inside of Γ the partial differential equation

$$(1 + q^2)r - 2pqs + (1 + p^2)t = 0,$$

where

$$p = z_x, \quad q = z_y, \quad r = z_{xx}, \quad s = z_{xy}, \quad t = z_{yy}.$$

IV.2. It has been observed (III.17) that this problem, in general, has no solution. The methods developed for the solution of the problem therefore necessarily are based on some further assumptions concerning the prescribed boundary conditions. In the work we are going to review in this chapter, it is supposed that

1. the curve Γ which bears the given boundary values is convex, and
2. the boundary function $\varphi(P)$ satisfies a *three-point condition*.

The three-point condition may be stated as follows. Let Γ^* be the JORDAN curve defined in the xyz-space by the equation $z = \varphi(P)$. Let P_1^*, P_2^*, P_3^* be three distinct points on Γ^*, and denote by Θ the positive acute angle between the xy-plane and the plane passing through P_1^*, P_2^*, P_3^*. If, for all possible positions of the points P_1^*, P_2^*, P_3^*, the quantity tg Θ is less than or equal to some fixed finite constant Δ, then we shall say that the boundary function $\varphi(P)$ satisfies the three-point condition with the constant Δ.

The three-point condition obviously implies that no three distinct points P_1^*, P_2^*, P_3^* of Γ^* are on the same vertical plane (the xy-plane being thought of as horizontal). This again implies that the curve Γ, bearing the given boundary values, is convex in the strict sense. That is to say, Γ must be a convex curve no arc of which reduces to a straight segment. Obviously, however, the three-point condition requires more than that; the condition implies a restriction not only upon Γ but also upon the boundary function $\varphi(P)$ given on Γ.

If Γ is not convex, then the boundary value problem generally is not solvable (see III.17). In this sense, the convexity of Γ is a necessary restriction. On the other hand, it will follow from the results obtained for the problem of PLATEAU in the parametric form that the three-point condition can be dropped (see VI.18). In the work concerned with the non-parametric problem the three-point condition is however absolutely essential.

IV.3. The three-point condition first appears in a short announcement which HILBERT gave of his method to justify the principle of DIRICHLET[1]. The ideas sketched in that announcement have been applied by LEBESGUE, in his Thesis[2], to develop a first attack on the problem of PLATEAU in the non-parametric form.

LEBESGUE considers the following problem. Let there be given, on a convex curve Γ in the xy-plane, a continuous function $\varphi(P)$ which satisfies the three-point condition with a constant Δ. Consider all the functions $z(x, y)$ which are continuous in and on Γ, and which reduce on Γ to the given function $\varphi(P)$. Determine in this class a function $z(x, y)$ such that the area $\mathfrak{A}(z)$ of the surface $z = z(x, y)$ be a minimum. $\mathfrak{A}(z)$ denotes here the area in the sense of LEBESGUE (see I.10).

LEBESGUE proceeds as follows. Consider first functions $z(x, y)$ such that the surface $z = z(x, y)$ is a polyhedron with a given number n of vertices and such that the boundary polygon of this polyhedron is inscribed in the curve Γ^* with equation $z = \varphi(P)$, where $\varphi(P)$ is the given boundary function. For a given n, the area $\mathfrak{A}(z)$ of this polyhedron is a continuous function of a finite number of parameters, and hence, for a given n, the problem $\mathfrak{A}(z) =$ minimum is an ordinary minimum problem and it has consequently a solution $z = z_n(x, y)$. It follows then from the definition of the area, in the sense of LEBESGUE, that $\mathfrak{A}(z_n)$ converges toward the greatest lower bound m of the areas $\mathfrak{A}(z)$ of all the surfaces $z = z(x, y)$ such that $z(x, y)$ is continuous in and on Γ, and reduces to $\varphi(P)$ on Γ.

Suppose then that the sequence $z_n(x, y)$ contains a uniformly convergent subsequence $z_{n_k}(x, y)$. Denote by $z(x, y)$ the limit function. Then $z(x, y)$ is continuous in and on Γ, and reduces to $\varphi(P)$ on Γ. Hence $m \leq \mathfrak{A}(z)$. On the other hand, on account of the lower semi-continuity of the area, $\mathfrak{A}(z) \leq \lim \mathfrak{A}(z_{n_k}) = m$. Consequently $\mathfrak{A}(z) = m$, that is to say $z(x, y)$ solves the proposed problem $\mathfrak{A}(z) =$ minimum.

The point therefore is to show that the sequence $z_n(x, y)$ contains a uniformly convergent subsequence. This will be established if it is shown that the functions $z_n(x, y)$ satisfy the LIPSCHITZ condition with the same constant, on account of a well-known theorem of ARZELÀ.

---

[1] HILBERT: Über das DIRICHLETsche Prinzip. Jber. Deutsch. Math.-Vereinig. Vol. 8 (1900) pp. 184—188.

[2] Intégrale, longueur, aire. Ann. Mat. pura appl. Vol. 7 (1920) pp. 231—359.

LEBESGUE shows then that the functions $z_n(x, y)$ satisfy the LIP-SCHITZ condition with the constant Δ of the three-point condition. He reasons as follows. The polyhedron $z = z_n(x, y)$ has no convex vertices; otherwise it would be possible to lessen its area without increasing the number of its vertices, in contradiction with its minimizing property. Since the polyhedron has no convex vertices, no plane can intersect it in a closed polygon. From this it follows that every plane which contains one of the faces of the polyhedron must intersect the curve Γ^*, in which the boundary polygon is inscribed, in at least three distinct points. Hence, on account of the three-point condition, the tangent of the positive acute angle between the xy-plane and any face of the polyhedron is less than or equal to the constant Δ of the three-point condition. This obviously implies that $z_n(x, y)$ satisfies the LIPSCHITZ condition with the constant Δ.

IV.4. This reasoning is essentially the same as that sketched by HILBERT for the DIRICHLET problem in the announcement mentioned above. We do not insist on details since every step of the reasoning will be generalized and discussed in connection with later investigations. It is however important to evaluate the implications of the result of LEBESGUE for the problem of PLATEAU.

If the solution $z(x, y)$ of the problem $\mathfrak{A}(z) =$ minimum would be sufficiently regular, the area $\mathfrak{A}(z)$ of the surface $z = z(x, y)$ would be given by the classical integral

$$\iint (1 + p^2 + q^2)^{\frac{1}{2}}\, dx\, dy.$$

The function $z(x, y)$ would then be a solution, with given boundary values, of the variation problem

$$\iint (1 + p^2 + q^2)^{\frac{1}{2}}\, dx\, dy = \text{minimum}.$$

The classical methods of the Calculus of Variations would then show that $z(x, y)$ satisfies the partial differential equation

$$(1 + q^2)r - 2pqs + (1 + p^2)t = 0.$$

There remains therefore the problem of showing that the solution $z = z(x, y)$ of the problem $\mathfrak{A}(z) =$ minimum has the differential coefficients necessary for the application of the classical methods of the Calculus of Variations. This is the sense in which LEBESGUE interpreted his result as a preliminary step toward the solution of the problem of PLATEAU. We are going to review presently the investigations which resulted in the solution of the problem left open by LEBESGUE.

IV.5. The problem clearly consists of handling the solution of a two-dimensional variation problem under much more adverse conditions than those considered in the classical Calculus of Variations. In the case of one-dimensional variation problems, the well-known lemma of DU BOIS REYMOND may be considered as a step in this direction. The generalization of this lemma for two-dimensional variation problems has been obtained by HAAR[1] and will play an important part in the sequel.

Lemma of HAAR. Given, in a JORDAN region R of the xy-plane, two continuous[2] functions $u(x, y)$, $v(x, y)$ such that

$$\iint_R (u\zeta_x + v\zeta_y)\,dx\,dy = 0$$

for every function $\zeta(x, y)$ which is continuous in R, vanishes on the boundary of R, and which has continuous partial derivatives of the first order in the interior of R. Then there exists, in the interior of R, a single-valued function $\omega(x, y)$ such that

$$\omega_y = u, \quad \omega_x = -v.$$

The point to the lemma of HAAR is that no assumptions are made concerning the existence of the partial derivatives of $u(x, y)$, $v(x, y)$. Indeed, if we suppose that $u(x, y)$, $v(x, y)$ have continuous first partial derivatives, the lemma of HAAR reduces to the classical argument used in Calculus of Variations to discuss the first variation of a double integral[3].

IV.6. The original proof of HAAR has been simplified by LICHTEN-STEIN[4], SCHAUDER[5] and by HAAR[6] himself. Some of the arguments used in these proofs deserve separate consideration.

a) *The one-dimensional lemma.* If $f(x)$ is continuous in a closed interval $a \leq x \leq b$ and is such that

$$\int_a^b f(x)\,\varphi'(x)\,dx = 0$$

for every $\varphi(x)$ which is continuous in $a \leq x \leq b$, vanishes for $x = a$ and $x = b$, and which has a continuous first differential coefficient in $a < x < b$, then $f(x)$ is constant.

The following proof[7] apparently is not generally known. Let $\varphi(x)$ have the properties described above and denote by γ any constant. Then

$$\int_a^b (f(x) - \gamma)\,\varphi'(x)\,dx = \int_a^b f(x)\,\varphi'(x)\,dx - \gamma \int_a^b \varphi'(x)\,dx = 0 - 0 = 0. \quad (4.1)$$

[1] Über die Variation der Doppelintegrale. J. reine angew. Math. Vol. 149 (1919) pp. 1—18.

[2] Later on HAAR generalized this lemma in several ways. See the expository presentation by A. HAAR: Zur Variationsrechnung. Abh. math. Semin. Hamburg. Univ. Vol. 8 (1930).

[3] See for instance BOLZA: Vorlesungen über Variationsrechnung, pp. 653—655.

[4] Bemerkungen über das Prinzip der virtuellen Verrückungen. Ann. Soc. Polon. math. (1924).

[5] Über die Umkehrung eines Satzes aus der Variationsrechnung. Acta Litt. Sci. Szeged Vol. 4 (1929) pp. 38—50.

[6] Zur Variationsrechnung. Abh. math. Semin. Hamburg. Univ. Vol. 8 (1930).

[7] See G. A. BLISS: Calculus of Variations (No. 1 of the Carus Mathematical Monographs).

Choose

$$\gamma = \frac{1}{b-a} \int_a^b f(x)\, dx.$$

Then

$$\varphi(x) = \int_a^x (f(\xi) - \gamma)\, d\xi$$

obviously has the desired properties. Since

$$\varphi'(x) = f(x) - \gamma$$

it follows from (4.1) that

$$\int_a^b (f(x) - \gamma)^2 dx = 0.$$

Hence $f(x) \equiv \gamma$.

b) *An integral test for complete differentials.* Given, in a closed rectangle $R: a \leq x \leq b,\ c \leq y \leq d$, two continuous functions $u(x, y)$, $v(x, y)$; the expression $u\,dy - v\,dx$ is called a complete differential if there exists, in the interior of R, a single-valued function $\omega(x, y)$ such that $\omega_x = -v$, $\omega_y = u$. If u, v have continuous derivatives of the first order in the interior of R, then $u\,dy - v\,dx$ is a complete differential if and only if $u_x + v_y \equiv 0$. This is the classical differential test. The vanishing of the line integral

$$\int u\,dy - v\,dx$$

for every closed, sufficiently good, curve in R is a classical integral test which remains valid under the single assumption that u, v are continuous. This is the test used by HAAR in the original proof of his lemma.

SCHAUDER[1], in his proof of the lemma of HAAR, used another interesting integral test. *Put*

$$U(x, y) = \int_c^y u(x, \eta)\, d\eta, \quad V(x, y) = \int_a^x v(\xi, y)\, d\xi. \tag{4.2}$$

Then $u\,dy - v\,dx$ is a complete differential if and only if there exist two functions $\varphi(x)$, $\psi(y)$ of the single variables x, y such that

$$U(x, y) + V(x, y) = \varphi(x) + \psi(y).$$

Indeed, suppose that $u\,dy - v\,dx$ is a complete differential: $u\,dy - v\,dx = d\omega$. Then obviously

$$U(x, y) = \omega(x, y) - \omega(x, c),$$
$$V(x, y) = \omega(a, y) - \omega(x, y),$$

and consequently $U(x, y) + V(x, y) = \omega(a, y) - \omega(x, c)$.

Suppose secondly that

$$U(x, y) + V(x, y) = \varphi(x) + \psi(y),$$

and put $\omega(x, y) = U(x, y) - \varphi(x) = -V(x, y) + \psi(y)$.

Then

$$\omega_x = -V_x = -v, \quad \omega_y = U_y = u.$$

[1] Über die Umkehrung eines Satzes aus der Variationsrechnung. Acta Litt. Sci. Szeged Vol. 4 (1929) pp. 38–50.

IV.7. The lemma of HAAR now may be proved as follows[1]. First it is clear that it is sufficient to prove the lemma for a rectangle $R : a \leq x \leq b, c \leq y \leq d$. Choose any function $l(x)$ which is continuous in $a \leq x \leq b$, vanishes for $x = a$ and $x = b$, and which has a continuous differential coefficient $l'(x)$ for $a < x < b$. Choose any function $m(y)$ having the same properties with respect to the interval $c \leq y \leq d$. Then $\zeta(x, y) = l(x) m(y)$ has all the properties required in the statement of the lemma of HAAR, and thus

$$\int_a^b \int_c^d (u l' m + v l m') \, dx \, dy = 0.$$

Partial integration gives

$$\int_a^b \int_c^d H l' m' \, dx \, dy = 0, \tag{4.3}$$

where $H = U + V$, and U, V are given by (4.2). It follows then from (4.3), on account of the one-dimensional lemma (see IV.6), that

$$\int_c^d H(x, y) m'(y) \, dy$$

is constant, that is to say independent of x. Hence

$$\int_c^d H(x, y) m'(y) \, dy = \int_c^d H(a, y) m'(y) \, dy,$$

or

$$\int_c^d [H(x, y) - H(a, y)] m'(y) \, dy = 0.$$

Applying again the one-dimensional lemma, we see that $H(x, y) - H(a, y)$ is constant for every fixed value of x. Hence

$$H(x, y) - H(a, y) = H(x, c) - H(a, c) = H(x, c), \tag{4.4}$$

since obviously $H(a, c) = 0$. (4.4) gives

$$H = U + V = H(a, y) + H(x, c).$$

It follows then from the integral test of SCHAUDER that $u \, dy - v \, dx$ is a complete differential, as asserted by the lemma of HAAR.

IV.8. A. HAAR applied his lemma to variation problems of the form $\int \int F(p, q) \, dx \, dy = $ minimum in the following way[2]. Consider, in a JORDAN region R, all the functions $z(x, y)$ which are continuous in R, which reduce on the boundary of R to a given boundary function, and which have continuous partial derivatives $z_x = p$, $z_y = q$ in the interior

---

[1] J. SCHAUDER: Über die Umkehrung eines Satzes aus der Variationsrechnung. Acta Litt. Sci. Szeged Vol. 4 (1929) pp. 38—50. — A. HAAR: Zur Variationsrechnung. Abh. math. Semin. Hamburg. Univ. Vol. 8 (1930).

[2] A. HAAR: Über die Variation der Doppelintegrale. J. reine angew. Math. Vol. 149 (1919) pp. 1—18.

of R. Suppose there exists in this class a function $z(x, y)$ which solves the variation problem

$$\iint_R F(p, q)\, dx\, dy = \text{minimum}.$$

Let $\zeta(x, y)$ denote any function which is continuous in R, vanishes on the boundary of R, and which has continuous partial derivatives of the first order in the interior of R. Let ε be a small parameter and put

$$J(\varepsilon) = \iint_R F(p + \varepsilon \zeta_x, q + \varepsilon \zeta_y)\, dx\, dy.$$

On account of the minimizing property of $z(x, y)$, this function $J(\varepsilon)$ has a minimum for $\varepsilon = 0$. Hence

$$J'(0) = \iint_R (F_p \zeta_x + F_q \zeta_y)\, dx\, dy = 0.$$

There follows then from the lemma of HAAR the existence, in the interior of R, of a single-valued function $\omega(x, y)$ such that

$$F_p = \omega_y, \qquad F_q = -\omega_x. \tag{4.5}$$

If the minimizing function $z(x, y)$ were known to have continuous partial derivatives of the second order also, then it would follow from (4.5) that

$$\frac{\partial}{\partial x} F_p + \frac{\partial}{\partial y} F_q = F_{pp} r + 2 F_{pq} s + F_{qq} t = 0. \tag{4.6}$$

This is the classical EULER-LAGRANGE equation of the problem. Examples show however that the minimizing function need not have partial derivatives of the second order. The lemma of HAAR permits us to derive the differential equations (4.5) under the single assumption of the existence and continuity of the partial derivatives of the first order.

IV.9. A. HAAR proposed the problem of applying the equations (4.5) to the study of the analytic character of the solutions of positive regular analytic variation problems of the form $\int\int F(p, q)\, dx\, dy = \text{minimum}$. If $F(p, q)$ is an analytic function of p, q and if it satisfies the inequalities

$$F_{pp} > 0, \quad F_{qq} > 0, \quad F_{pp} F_{qq} - F_{pq}^2 > 0 \tag{4.7}$$

for all values of p, q, then the problem $\int\int F(p, q)\, dx\, dy = \text{minimum}$ is called positive regular and analytic. If $z(x, y)$ is a solution, with continuous partial derivatives of the second order, of such a problem, then $z(x, y)$ satisfies the EULER-LAGRANGE equation (4.6). This is a quasi-linear equation of the second order, and it is of the elliptic type, on account of (4.7). Every solution of such an equation is analytic as soon as it has continuous partial derivatives of the first and second order[1]. Hence every solution of a regular analytic variation problem $\int\int F(p, q)\, dx\, dy = \text{minimum}$ is analytic as soon as it has continuous

---

[1] See L. LICHTENSTEIN: Neuere Entwicklung usw. Enzyklopädie der math. Wiss. Vol. 2 (3) pp. 1277—1334.

partial derivatives of the first and second order. The problem proposed by HAAR requires the obtaining of this conclusion under the assumption of the existence and continuity of the first partial derivatives only.

At the present time the solution of this problem is known only in two special cases, namely for the DIRICHLET problem

$$\frac{1}{2}\iint (p^2 + q^2)\, dx\, dy = \text{minimum},$$

and for the area-problem

$$\iint (1 + p^2 + q^2)^{\frac{1}{2}}\, dx\, dy = \text{minimum}.$$

In the case of the DIRICHLET problem, the equations of HAAR reduce to

$$p = z_x = \omega_y, \quad q = z_y = -\omega_x,$$

that is to say to the CAUCHY-RIEMANN equations. The fact that $z(x, y)$ is analytic follows therefore from classical theorems in the theory of functions of a complex variable.

We are going to review now the case $F = (1 + p^2 + q^2)^{\frac{1}{2}}$.

IV.10. We consider first the area problem in the parametric form[1]. Let $S : \mathfrak{x} = \mathfrak{x}(u, v)$, $u^2 + v^2 \leq 1$, be a regular surface of class C' (see II.2) bounded by a given curve, and suppose that the area of S is a minimum.

Consider an interior point P_0 of S. Choose the coordinate system xyz in such a way that none of the coordinate planes is parallel to the normal of S at P_0. Denote by S_0 a small simply connected portion of S comprising P_0. If S_0 is sufficiently small, then S_0 can be represented in any one of the three forms $z = z(x, y)$, $x = x(y, z)$, $y = y(z, x)$, where $z(x, y), x(y, z), y(z, x)$ are single-valued functions with continuous first partial derivatives.

Denote by X, Y, Z the direction-cosines of the normal to the surface. The first step is to prove that the three expressions

$$Y\,dz - Z\,dy, \quad Z\,dx - X\,dz, \quad X\,dy - Y\,dx$$

are complete differentials on S_0.[2]

Consider the representation $z = z(x, y)$ of S_0. Then $z(x, y)$ can be considered as a solution of the variation problem

$$\iint (1 + p^2 + q^2)^{\frac{1}{2}}\, dx\, dy = \text{minimum}.$$

Since $z(x, y)$ has continuous first partial derivatives, the result of HAAR can be applied. The equations of HAAR reduce in the present case to

$$\frac{p}{(1 + p^2 + q^2)^{\frac{1}{2}}} = \omega_y, \quad \frac{q}{(1 + p^2 + q^2)^{\frac{1}{2}}} = -\omega_x.$$

---

[1] T. RADÓ: Über den analytischen Charakter der Minimalflächen. Math. Z. Vol. 24 (1925) pp. 321- 327.

[2] Cf. II.11.

In other words, the expression

$$\frac{p\,dy - q\,dx}{(1 + p^2 + q^2)^{\frac{1}{2}}}$$

is a complete differential[1]. On the other hand, we have

$$X = -\frac{p}{(1 + p^2 + q^2)^{\frac{1}{2}}}, \quad Y = -\frac{q}{(1 + p^2 + q^2)^{\frac{1}{2}}},$$

and consequently

$$X\,dy - Y\,dx = -\frac{p\,dy - q\,dx}{(1 + p^2 + q^2)^{\frac{1}{2}}}.$$

This proves that $X\,dy - Y\,dx$ is a complete differential. Using the representations $x = x(y, z)$, $y = y(z, x)$ of S_0, we see that $Y\,dz - Z\,dy$, $Z\,dx - X\,dz$ are also complete differentials.

Consider then again the representation $z = z(x, y)$ of S_0. It can be supposed that the xy-projection of S_0 is a rectangle $R : x_1 \leqq x \leqq x_2$, $y_1 \leqq y \leqq y_2$. We have for the components X, Y, Z of the unit normal vector

$$X = -\frac{p}{W}, \quad Y = -\frac{q}{W}, \quad Z = \frac{1}{W}, \quad W = (1 + p^2 + q^2)^{\frac{1}{2}}.$$

The preceding result is then expressed by the equations

$$\left.\begin{aligned}
&-\frac{pq}{W}\,dx - \frac{1 + q^2}{W}\,dy = d\omega_1, \quad \frac{1 + p^2}{W}\,dx + \frac{pq}{W}\,dy = d\omega_2, \\
&\frac{q}{W}\,dx - \frac{p}{W}\,dy = d\omega_3,
\end{aligned}\right\} \tag{4.8}$$

where ω_1, ω_2, ω_3 denote three auxiliary functions which are single-valued in R. By assumption p, q and consequently the first partial derivatives of ω_1, ω_2, ω_3 are continuous. Introduce new parameters α, β by the equations

$$\alpha = x, \quad \beta = \omega_1(x, y).$$

From $\beta_y = \omega_{1y} = -\frac{1 + q^2}{W} < 0$ it follows then that R is carried in a one-to-one way into a certain region R^* of the $\alpha\beta$-plane. Direct computation shows then (cf. II.16) that x and ω_1, y and ω_2, z and ω_3, as functions of α and β, satisfy in R^* the CAUCHY-RIEMANN equations

$$x_\alpha = \omega_{1\beta}, \quad x_\beta = -\omega_{1\alpha},$$
$$y_\alpha = \omega_{2\beta}, \quad y_\beta = -\omega_{2\alpha},$$
$$z_\alpha = \omega_{3\beta}, \quad z_\beta = -\omega_{3\alpha}.$$

Since x, y, z, ω_1, ω_2, ω_3 as functions of α, β have continuous first partial derivatives, it follows from classical theorems that the six functions involved are analytic functions of α and β. Since

$$\frac{\partial(\alpha, \beta)}{\partial(x, y)} = -\frac{1 + q^2}{W} < 0,$$

---

[1] Cf. II.6.

it follows from well-known theorems on implicit functions that α, β are analytic functions of x and y. As z is an analytic function of α, β and as α, β are analytic functions of x and y, it follows that z is an analytic function of x and y. Thus we have the theorem[1]:

If a regular surface S of class C', bounded by a given curve, has a minimum area, then S is analytic (and consequently is a minimal surface).

IV.11. The minimizing property of S has been used only to establish the equations (4.8); from that point on, only those equations were used. Hence we have the following theorem[2].

Let $z(x, y)$ be a function which has continuous partial derivatives of the first order in a JORDAN region R. Suppose there exist in R three single-valued functions $\omega_1, \omega_2, \omega_3$ such that

$$\omega_{1x} = -\frac{pq}{W}, \qquad \omega_{1y} = -\frac{1 + q^2}{W},$$

$$\omega_{2x} = \frac{1 + p^2}{W}, \qquad \omega_{2y} = \frac{pq}{W},$$

$$\omega_{3x} = \frac{q}{W}, \qquad \omega_{3y} = -\frac{p}{W}.$$

Then $z(x, y)$ is an analytic function of x, y and satisfies the partial differential equation

$$(1 + q^2)r - 2pqs + (1 + p^2)t = 0.$$

The second half of the statement follows of course from the first half directly by differentiation.

IV.12. Consider now *the area problem in the non-parametric form*

$$\iint (1 + p^2 + q^2)^{\frac{1}{2}} dx dy = \text{minimum}.$$

Suppose we have, in a JORDAN region R, a solution $z(x, y)$ with given boundary values and having continuous first partial derivatives in the interior of R. Using variations of the form $z + \varepsilon \zeta$, we obtain from the lemma of HAAR the result that $X dy - Y dx$ is a complete differential. The surface $z = z(x, y)$ is now only known to have a smallest area with respect to surfaces of the same form. The reasoning which showed, in the case of the parametric problem, that $Y dz - Z dy$, $Z dx - X dz$ are also complete differentials is therefore not legitimate in the present case. Still, the result remains valid[2]. The situation can be handled by the so-called *method of the variation of the independent variables.*

[1] T. RADÓ: Über den analytischen Charakter der Minimalflächen. Math. Z. Vol. 24 (1925) pp. 321—327.

[2] T. RADÓ: Bemerkung über die Differentialgleichungen zweidimensionale. Variationsprobleme. Acta Litt. Sci. Szeged Vol. 3 (1925) pp. 147—156.

Let us consider the general problem

$$\iint F(p, q) \, dx \, dy = \text{minimum}.$$

Let $z(x, y)$ be a solution with given boundary values, and suppose only that $z(x, y)$ has continuous first partial derivatives. It clearly is legitimate to suppose that the region R in which the problem is considered is a circle K. Denote by $t(x, y)$ a function continuous in K, vanishing on the boundary of K, and having continuous and bounded first partial derivatives in K. Let ε be a small parameter and define a transformation $T(\varepsilon)$ by the formulas

$$T(\varepsilon) : x = u + \varepsilon t(u, v), \quad y = v, \quad (u, v) \text{ in } K.$$

For sufficiently small values of ε this is a one-to-one and continuous transformation of K into itself. Hence, u, v can be expressed as single-valued functions of x, y and ε. It follows in this way that the equations

$$x = u + \varepsilon t(u, v), \quad y = v, \quad z = z(u, v), \quad (u, v) \text{ in } K,$$

where z is the minimizing function under investigation, define a family of admissible surfaces. For these surfaces the integral $\iint F(p, q) \, dx \, dy$ becomes a function $J(\varepsilon)$ which has a minimum for $\varepsilon = 0$. Hence $J'(0) = 0$. After suitable transformations, the function $J(\varepsilon)$ appears in the form

$$J(\varepsilon) = \iint_K F\left(\frac{p}{1 + \varepsilon t_x}, \; q - \frac{\varepsilon p \, t_y}{1 + \varepsilon t_x}\right)(1 + \varepsilon t_x) \, dx \, dy.$$

The condition $J'(0) = 0$ gives therefore the equation

$$\iint_K [(F - p F_p) t_x - p F_q t_y] \, dx \, dy = 0.$$

It follows then from the lemma of HAAR that

$$p F_q \, dx + (F - p F_p) \, dy$$

is a complete differential. Using in a similar way variations of the independent variable y, we obtain the result that

$$(F - q F_q) \, dx + q F_p \, dy$$

is a complete differential. Finally it follows by means of the usual variation $z + \varepsilon \zeta$ that $F_q \, dx - F_p \, dy$ is a complete differential. That is to say:

If $z(x, y)$ is a solution with continuous first partial derivatives of the boundary value problem for the variation problem

$$\iint F(p, q) \, dx \, dy = \text{minimum},$$

then the three expressions

$$p F_q \, dx + (F - p F_p) \, dy, \quad (F - q F_q) \, dx + q F_p \, dy, \quad F_q \, dx - F_p \, dy$$

are complete differentials. In other words: there exist three single-valued auxiliary functions $\omega_1(x, y)$, $\omega_2(x, y)$, $\omega_3(x, y)$ which satisfy, together with the minimizing function $z(x, y)$, the following system of partial differential equations[1],

$$\left.\begin{array}{ll} \omega_{1x} = pF_q, & \omega_{1y} = F - pF_p, \\ \omega_{2x} = F - qF_q, & \omega_{2y} = qF_p, \\ \omega_{3x} = F_q, & \omega_{3y} = -F_p. \end{array}\right\} \quad (4.9)$$

The last two equations are those obtained by HAAR himself. In case the minimizing function $z(x, y)$ is known to have continuous partial derivatives of the second order, the auxiliary functions ω_1, ω_2, ω_3 can be eliminated by crosswise differentiation and there results the classical EULER-LAGRANGE equation. Conversely, it is immediate that if $z(x, y)$ satisfies the EULER-LAGRANGE equation, then the auxiliary functions ω_1, ω_2, ω_3 exist. Indeed, the EULER-LAGRANGE equation is the common integrability condition for the three couples of equations comprised in the above system. The point is again that the equations (4.9) can be established under the single assumption that the minimizing function has continuous partial derivatives of the first order.

IV.13. In the case $F = (1 + p^2 + q^2)^{\frac{1}{2}}$ the equations (4.9) reduce to those considered in IV.11. Hence, on account of the remark made there, we have the following theorem[1].

If $z(x, y)$ is a solution, with continuous partial derivatives of the first order, of the boundary value problem for the variation problem

$$\int\int (1 + p^2 + q^2)^{\frac{1}{2}} dx dy,$$

then $z(x, y)$ is analytic and satisfies the partial differential equation

$$(1 + q^2)r - 2pqs + (1 + p^2)t = 0. \quad (4.10)$$

IV.14. We are going to review at present a paper of A. HAAR[2] in which he gave important generalizations and applications of the results discussed so far in this Chapter. In § 2 and § 3 of his paper, HAAR generalizes the theorem of IV.13 for the case when the minimizing function $z(x, y)$ is only known to satisfy the LIPSCHITZ condition. If a function $z(x, y)$ satisfies the condition of LIPSCHITZ in a domain D, then $p = z_x$ and $q = z_y$ exist almost everywhere in D and are bounded and measurable functions[3]. The expression $(1 + p^2 + q^2)^{\frac{1}{2}}$ is then also a bounded and measurable function of x, y and hence the integral

$$\int\int_D (1 + p^2 + q^2)^{\frac{1}{2}} dx dy$$

[1] T. RADÓ: Bemerkung über die Differentialgleichungen zweidimensionaler Variationsprobleme. Acta Litt. Sci. Szeged Vol. 3 (1925) pp. 147—156.

[2] A. HAAR: Über das PLATEAUsche Problem. Math. Ann. Vol. 97 (1927) pp. 124—258.

[3] The theory of functions of two variables, satisfying the LIPSCHITZ condition, has been the object of important investigations of H. RADEMACHER: Über partielle und totale Differenzierbarkeit I, II. Math. Ann. Vol. 79 (1919) pp. 340—359 and Vol. 81 (1920) pp. 52—63.

exists in the LEBESGUE sense. Thus there is a good sense to the statement of the variation problem

$$\iint_D (1 + p^2 + q^2)^{\frac{1}{2}} dx dy = \text{minimum} \qquad (4.11)$$

for the class of functions $z(x, y)$ which satisfy the LIPSCHITZ condition. Using the modern theory of functions of real variables, HAAR verifies that all the arguments which have been used to prove the theorem of IV.13 remain valid under the single assumption that the minimizing function $z(x, y)$ satisfies the LIPSCHITZ condition. He obtains in this way the theorem:

If $z(x, y)$ is a solution, satisfying the condition of LIPSCHITZ, of the boundary value problem for the variation problem (4.11), then $z(x, y)$ is analytic and satisfies the partial differential equation (4.10).

IV.15. HAAR observes that this result can be applied to the existence theorem obtained by LEBESGUE (see IV.3). Indeed, GEÖCZE proved (see I.13) that if $z(x, y)$ satisfies the condition of LIPSCHITZ, then the area $\mathfrak{A}(z)$ of the surface $z = z(x, y)$ is given by the classical integral

$$\iint (1 + p^2 + q^2)^{\frac{1}{2}} dx dy. \qquad (4.12)$$

Hence the result of LEBESGUE concerning the problem $\mathfrak{A}(z) = \text{minimum}$ (see IV.3) implies the following existence theorem.

Let there be given, on a convex JORDAN curve Γ in the xy-plane, a function $\varphi(P)$ of the point P varying on Γ which satisfies the three-point condition with some constant Δ. Consider all the functions $z(x, y)$ which satisfy, in the JORDAN region bounded by Γ, the LIPSCHITZ condition and which reduce on Γ to the given function $\varphi(P)$. Then there exists in this class a function $z_0(x, y)$ which minimizes the integral (4.12) extended over the interior of Γ.

IV.16. It follows then from the theorem in IV.14 that this function $z_0(x, y)$ is analytic and satisfies the partial differential equation (4.10). Hence the theorem:

Let there be given, on a convex JORDAN curve Γ, a function $\varphi(P)$ which satisfies the three-point condition with some constant Δ. Then there exists, in the JORDAN region bounded by Γ, a solution of the partial differential equation

$$(1 + q^2) r - 2pqs + (1 + p^2) t = 0,$$

which reduces to $\varphi(P)$ on Γ.

IV.17. In § 1 of his paper under discussion, HAAR generalizes the existence theorem of IV.15 for positive regular variation problems of the form

$$\iint F(p, q) dx dy = \text{minimum}.$$

Such a problem is called positive regular if F satisfies the inequalities

$$F_{pp} > 0, \quad F_{qq} > 0, \quad F_{pp} F_{qq} - F_{pq}^2 > 0 \qquad (4.13)$$

for all values of p and q. The existence proof presented by HAAR for this general case is considerably simpler than the existence proof which follows from the results of LEBESGUE and GEÖCZE for the special case $F = (1 + p^2 + q^2)^{\frac{1}{2}}$. The reviewer believes that this justifies the detailed discussion of the general case in this report. We shall first consider the main arguments used by HAAR and we shall then describe the existence proof itself.

IV.18. Let $z(x, y)$ be a function continuous in a JORDAN region bounded by a convex JORDAN curve Γ. Suppose that the boundary values of $z(x, y)$ satisfy the three-point condition with some constant Δ. Suppose also that $z(x, y)$ has the property that there exists no plane which intersects the surface $z = z(x, y)$ in a closed curve (this assumption will be stated in a more exact form presently). One of the main arguments of HAAR is then the theorem that under these circumstances the function $z(x, y)$ satisfies the LIPSCHITZ condition with the constant Δ of the three-point condition.

Similar arguments, concerned with various special types of surfaces $z = z(x, y)$, played an important part in most of the investigations on the problem of PLATEAU in the non-parametric form[1]. For instance, the method of LEBESGUE reviewed in IV.3 is based on the special case concerned with polyhedrons. The general theorem has been stated by HAAR in the course of preliminary investigations on the problem of PLATEAU and has been proved first by T. RADÓ[2]. A much simpler proof has then been given by J. v. NEUMANN[3].

The exact statement of the theorem is based on the following definitions. Let $f(x, y)$ be continuous in a JORDAN region R. Let D be any domain (connected open set) comprised in R. Denote by D^* the set of all boundary points of D, and by M^*, m^* the maximum and minimum respectively of $f(x, y)$ on D^*. The function $f(x, y)$ is called *monotonic*[4] in the JORDAN region R if the condition

$$m^* \leqq f(x, y) \leqq M^* \text{ for } (x, y) \text{ in } D$$

is satisfied for every domain D in R.

Let a, b, c be three constants. If the function $f(x, y) - (ax + by + c)$ is monotonic in R for every choice of the constants a, b, c, then $f(x, y)$ is called a *saddle-function* in R. Using this terminology, we may state the theorem as follows.

---

[1] See for references A. HAAR: Über das PLATEAUsche Problem. Math. Ann. Vol. 97 (1927) p. 127 and 141.

[2] T. RADÓ: Geometrische Betrachtungen über zweidimensionale reguläre Variationsprobleme. Acta Litt. Sci. Szeged Vol. 2 (1926) pp. 228—253.

[3] J. v. NEUMANN: Über einen Satz der Variationsrechnung. Abh. math. Sem. Hamburg. Univ. Vol. 8 (1931) pp. 28—31.

[4] H. LEBESGUE: Intégrale, longueur, aire. Ann. Mat. pura appl. Vol. 7 (1902) pp. 231—359.

If $z(x, y)$ is a saddle-function in a convex JORDAN region R, and if the boundary values of $z(x, y)$ satisfy the three-point condition with some constant Δ, then $z(x, y)$ satisfies in R the LIPSCHITZ condition with the constant Δ.

IV.19. The proof of this theorem, although quite elementary, is rather involved even in the simplified form due to J. v. NEUMANN, and therefore we restrict ourselves to the following remarks. Let (x_1, y_1), (x_2, y_2) be any two distinct points in R; the proof consists of showing that the inequality

$$\frac{|z(x_2, y_2) - z(x_1, y_1)|}{[(x_2 - x_1)^2 + (y_2 - y_1)^2]^{\frac{1}{2}}} > \Delta \qquad (4.14)$$

is impossible. This is practically obvious if at least one of the two points is on the boundary of R^1. The complications arise solely in the general case when both points are interior points of R.

T. RADÓ observed that as far as the existence proof of HAAR is concerned, these complications easily can be avoided[1]. Indeed, the functions $z(x, y)$, to which the theorem will be applied in the course of the existence proof, will possess the following additional property. Let h, k be two constants. Denote by $R^{h,k}$ the JORDAN region obtained from the JORDAN region R by the transformation $\bar{x} = x + h, \bar{y} = y + k$. Then the function $z(x - h, y - k) - z(x, y)$ is monotonic in the region consisting of the common points of R and $R^{h,k}$, for every choice of the constants h and k.

Consider then two interior points (x_1, y_1), (x_2, y_2) of R such that (4.14) is satisfied. Put $h = x_2 - x_1$, $k = y_2 - y_1$. From (4.14) and from the monotonic character of $z(x - h, y - k) - z(x, y)$ it follows then immediately that (4.14) is satisfied also for a couple of points $(\bar{x}_1, \bar{y}_1)$, $(\bar{x}_2, \bar{y}_2)$, at least one of which is a boundary point of R. Hence it is sufficient to disprove (4.14) in the almost trivial case when at least one of the two points is on the boundary[1].

IV.20. Another important argument in the existence proof of HAAR is the following theorem.

Suppose that $F(p, q)$ satisfies, for all values of p and q, the inequalities (4.13). Consider then, in a JORDAN region R, a sequence $z_n(x, y)$ converging uniformly to a function $z(x, y)$. Suppose that all these functions satisfy in R the LIPSCHITZ condition (not necessarily with the same constant). Then

$$\varliminf \iint_R F(p_n, q_n) \, dx \, dy \geq \iint_R F(p, q) \, dx \, dy,$$

where $p_n = z_{nx}$, $q_n = z_{ny}$, $p = z_x$, $q = z_y$.

The theorem expresses the *lower semi-continuity* of the integral $\iint F(p, q) \, dx \, dy$. In the special case $F = (1 + p^2 + q^2)^{\frac{1}{2}}$, the integral

---

[1] T. RADÓ: Über zweidimensionale reguläre Variationsprobleme. Math. Ann. Vol. 101 (1929) pp. 620—632. See in particular § 1, No. 2.

is equal to the area of the surface $z = z(x, y)$ in the sense of Lebesgue (see I.13), and hence the theorem is a consequence of the lower semi-continuity of the area (see I.12). The lower semi-continuity of the area-integral has been generalized to various classes of simple and multiple integrals[1].

For the integral in which we are interested, a very simple proof is obtained by applying a device used by Haar[2] to the method which results from the work of Geöcze for the special case $(1 + p^2 + q^2)^{\frac{1}{2}}$. This proof proceeds in the following steps[3]. First observe that it is sufficient to consider the case when the Jordan region R is a square. Use the notation

$$J(z) = \iint_R F(p, q)\,dx\,dy,$$

where z is any function which satisfies the condition of Lipschitz in R. Subdivide R into m^2 congruent small squares, and put

$$\Sigma_m(z) = \sum_{k=1}^{m^2} F\left(\frac{1}{\sigma_{m,k}} \iint_{\sigma_{m,k}} p\,dx\,dy\,,\ \frac{1}{\sigma_{m,k}} \iint_{\sigma_{m,k}} q\,dx\,dy\right)\sigma_{m,k}\,.$$

In this formula, $\sigma_{m,1}$, $\sigma_{m,2}$, ..., σ_{m,m^2} denote the small squares into which R has been subdivided; $\sigma_{m,k}$ also denotes the area of $\sigma_{m,k}$. The following properties of $\Sigma_m(z)$ will be used.

1. $\Sigma_m(z) \to J(z)$ for $m \to \infty$.

2. $\Sigma_m(z) \leqq J(z)$.

3. If a sequence $z_n(x, y)$ converges uniformly to a function $z(x, y)$ in R, and if all these functions satisfy the condition of Lipschitz in R (not necessarily with the same constant), then

$$\lim_{n \to \infty} \Sigma_m(z_n) = \Sigma_m(z)$$

for every fixed value of m.

The proofs of 1 and 3 are immediate if all the functions concerned are supposed to have continuous first partial derivatives. The validity of the arguments used in this elementary case can then be extended on account of well-known general theorems of the theory of functions of real variables. As to 2, observe first that we have, on account of the inequalities (4.13),

$$F\left(\frac{a_1 + \cdots + a_\nu}{\nu}\,,\ \frac{b_1 + \cdots + b_\nu}{\nu}\right) \leqq \frac{F(a_1, b_1) + \cdots + F(a_\nu, b_\nu)}{\nu}$$

---

[1] See, also for references, L. Tonelli: Sur la semi-continuité des intégrales doubles du Calcul des Variations. Acta math. Vol. 53 (1929) pp. 325—346 and J. E. McShane: On the semi-continuity of double integrals. Ann. of Math. Vol. 33 (1932) pp. 460—484.

[2] Über das Plateausche Problem. Math. Ann. Vol. 97 (1927) pp. 124—258.

[3] T. Radó: Über zweidimensionale reguläre Variationsprobleme. Math. Ann. Vol. 101 (1929) pp. 620—632.

for every value of the positive integer ν and of the real constants $a_1, \ldots, a_\nu, b_1, \ldots, b_\nu$.[1] The inequality 2 follows then immediately by a passage to the limit. The main theorem

$$\varliminf J(z_n) \geqq J(z)$$

is then proved as follows. 2 gives $\Sigma_m(z_n) \leqq J(z_n)$. For $n \to \infty$ it follows, on account of 3, that $\Sigma_m(z) \leqq \varliminf J(z_n)$. For $m \to \infty$, it follows, on account of 1, that $J(z) \leqq \varliminf J(z_n)$.

IV.21. The statement of the existence theorem of HAAR is as follows.

Let there be given, on a convex curve Γ of the xy-plane, a continuous function $\varphi(P)$ of the point P varying on Γ which satisfies the three-point condition with some constant Δ. Consider the class of all functions $z(x, y)$ which satisfy the LIPSCHITZ condition in the JORDAN region R bounded by Γ and reduce to $\varphi(P)$ on Γ. Then there exists in this class a solution of the variation problem

$$\iint\limits_R F(p, q)\, dx\, dy = \text{minimum}, \tag{4.15}$$

if the problem is positive regular, that is to say if

$$F_{pp} > 0, \quad F_{qq} > 0, \quad F_{pp}F_{qq} - F_{pq}^2 > 0$$

for all values of p and q.

IV.22. The proof proceeds in the following steps. Denote by K the class of functions described above. Denote by K_L the class of those functions in K which satisfy the LIPSCHITZ condition with a given constant L. A first remark is then that for $L \geqq \Delta$ the class K_L certainly is not empty. HAAR observes that the harmonic function which reduces to $\varphi(P)$ on Γ satisfies the LIPSCHITZ condition with the constant Δ and consequently is comprised in K_L for $L \geqq \Delta$. A more elementary argument is obtained by the following construction. Denote by Γ^* the curve in the xyz-space with the equation $z = \varphi(P)$. Take a fixed point P_0^* on Γ^*. Then the straight segments connecting P_0^* with a variable point P^* of Γ^* constitute a surface S with the obvious property: if A^*, B^* are any two points of S, then there exists a plane through A^*, B^* which intersects Γ^* in at least three distinct points. If then $z = z(x, y)$ is the equation of S, it follows from the three-point condition that $z(x, y)$ satisfies the LIPSCHITZ condition with the constant Δ.

IV.23. Let L be any constant $> \Delta$. Consider the problem (4.15) for the class K_L only. Then it readily is seen that this restricted problem has a solution. First, the class K_L is not empty (see IV.22). Secondly, since $|p|$, $|q|$ are both $\leqq L$, and since $F(p, q)$ is continuous, the absolute value of the integral is uniformly bounded for all functions of

---

[1] See for instance PÓLYA-SZEGÖ: Aufgaben und Lehrsätze. Vol. 1 pp. 51–52, problems 70 and 71.

the class K_L. The greatest lower bound m_L of the integral, for all functions of the class K_L, is therefore finite. There exists, by definition, a sequence of functions $z_n(x, y)$, comprised in K_L, such that

$$\iint\limits_{R} F(p_n, q_n)\, dx\, dy \rightarrow m_L.$$

Since the functions $z_n(x, y)$ satisfy the LIPSCHITZ condition with the same constant L, the sequence $z_n(x, y)$ contains a uniformly convergent subsequence $z_{n_k}(x, y)$. The limit function $z_0(x, y)$ clearly belongs again to K_L, and thus

$$\iint\limits_{R} F(p_0, q_0)\, dx\, dy \geqq m_L.$$

On the other hand, on account of IV.20,

$$\iint\limits_{R} F(p_0, q_0)\, dx\, dy \leqq \varliminf \iint\limits_{R} F(p_{n_k}, q_{n_k})\, dx\, dy = m_L.$$

Hence

$$\iint\limits_{R} F(p_0, q_0)\, dx\, dy = m_L.$$

IV.24. The function $z_0(x, y)$, obtained in this way, satisfies the LIPSCHITZ condition with the constant L, since it is comprised in the class K_L. *The heart of the existence proof is the fact that $z_0(x, y)$ satisfies the LIPSCHITZ condition with the constant $\Delta < L$.*

On account of IV.18, it is sufficient to verify that $z_0(x, y)$ is a saddle-function. In the special case $F = (1 + p^2 + q^2)^{\frac{1}{2}}$, the fact is geometrically obvious. Indeed, if $z_0(x, y)$ is not a saddle-function, then we have a plane $z = ax + by + c$, such that the difference $z_0(x, y)$ $- (ax + by + c)$ vanishes on the boundary of a certain domain D, while it is, say, positive in D itself. Replace then $z_0(x, y)$ in D by $ax + by + c$. There results a new function $\bar{z}_0(x, y)$. No secant of the new surface $z = \bar{z}_0(x, y)$ is steeper than the corresponding secant of the old surface $z = z_0(x, y)$, while the area of the new surface clearly is less than the area of the old surface, in contradiction with the minimizing property of $z_0(x, y)$. The analytic justification of this geometric argument can be extended immediately to the problem with a general $F(p, q)$.

If one desires to take advantage of the remark in IV.19, then it is necessary to verify that $z_0(x, y)$ possesses the additional property stated there. The case $F = (1 + p^2 + q^2)^{\frac{1}{2}}$ permits again of obvious geometric justification, which also suggests the proof for the general case[1].

IV.25. Let now $\bar{z}(x, y)$ be any function of the class K, and let $z_0(x, y)$ be the minimizing function for the class K_L. Consider the function

$$z(x, y) = z_0(x, y) + \Theta(\bar{z}(x, y) - z_0(x, y)),$$

[1] T. RADÓ: Über zweidimensionale reguläre Variationsprobleme. Math. Ann. Vol. 101 (1929) pp. 620—632.

where Θ is a real parameter. For every given value of Θ, $z(x, y)$ clearly belongs to the class K. The integral

$$\iint\limits_{R} F(p, q)\, dx\, dy$$

becomes a function $J(\Theta)$ of Θ, and it follows from the inequalities (4.13) immediately that $J''(\Theta) \geq 0$. Hence $J(\Theta)$ is convex.

For small values of Θ, $z(x, y)$ satisfies the LIPSCHITZ condition with practically the same constant as $z_0(x, y)$. As $z_0(x, y)$ satisfies the LIPSCHITZ condition with the constant $\Delta < L$ (see IV.24), it follows that for small values of Θ the function $z(x, y)$ belongs to the class K_L. Hence, on account of the minimizing property of $z_0(x, y)$, $J(0) \leq J(\Theta)$ for small values of Θ. A local minimum of a convex function is however necessarily an absolute minimum. Hence $J(0) \leq J(1)$, that is to say

$$\iint\limits_{R} F(p_0, q_0)\, dx\, dy \leq \iint\limits_{R} F(\bar{p}, \bar{q})\, dx\, dy. \tag{4.16}$$

In other words: $z_0(x, y)$ minimizes the integral not only in the restricted class K_L, but also in the whole class K. This proves the existence theorem stated in IV.21.

IV.26. The reasoning used in IV.25 may be replaced by the following argument. Let $\bar{z}(x, y)$ satisfy the condition of LIPSCHITZ with some constant $\bar{L}$ (since $\bar{z}$ belongs to K, such a constant exists by assumption). If $\bar{L} \leq L$, then $\bar{z}$ belongs to K_L, and (4.16) is a direct consequence of the minimizing property of z_0. If $\bar{L} > L > \Delta$, then consider the class $K_{\bar{L}}$. Let $\bar{z}_0$ be the minimizing function for the class $K_{\bar{L}}$ ($\bar{z}_0$ exists for the same reason as z_0). Then, by the definition of $\bar{z}_0$,

$$\iint\limits_{R} F(\bar{p}_0, \bar{q}_0)\, dx\, dy \leq \iint\limits_{R} F(\bar{p}, \bar{q})\, dx\, dy. \tag{4.17}$$

But $\bar{z}_0$ satisfies the condition of LIPSCHITZ with the constant Δ (for the same reason as z_0). Hence $\bar{z}_0$ belongs to the class K_L, since $\Delta < L$. Consequently, on account of the minimizing property of z_0,

$$\iint\limits_{R} F(p_0, q_0)\, dx\, dy \leq \iint\limits_{R} F(\bar{p}_0, \bar{q}_0)\, dx\, dy. \tag{4.18}$$

(4.16) follows now from (4.17) and (4.18).

IV.27. Suppose $F(p, q)$ is an *analytic* function of p, q. The question naturally arises as to whether the solution $z(x, y)$ of the variation problem (4.15), obtained by HAAR, is analytic. From the existence proof it only follows that $z(x, y)$ satisfies the LIPSCHITZ condition. In the special case $F = (1 + p^2 + q^2)^{\frac{1}{2}}$, the proof of the analytic character of $z(x, y)$ depended essentially on computations suggested by differential geometry. HAAR gave a discussion of the extremals for the general case $F(p, q)$ with the purpose to generalize the main theorems of the differential geometry of minimal surfaces[1]. While the analogies are striking, his results did not permit as yet to establish the analytic character of the solution of the variation problem under the only assumption that the solution satisfies the LIPSCHITZ condition. Under the assumption that the solution has continuous first partial derivatives which satisfy the LIPSCHITZ-HÖLDER condition with some exponent λ such that $0 < \lambda < 1$, the analytic character has been proved by E. HOPF[2].

IV.28. The boundary value problem for the equation

$$(1 + q^2)r - 2pqs + (1 + p^2)t = 0$$

has been also treated in papers of A. KORN[3], CH. MÜNTZ[4], S. BERNSTEIN[5] by methods whose discussion is beyond the scope of this report. The geometrical results obtained by these authors are less general than those explicitly considered in the present review.

Chapter V.

The problem of PLATEAU in the parametric form.

V.1. The method of GARNIER[6], which we shall consider first, is concerned with problem P_3 (see III.5), and permits us to carry out a program outlined by WEIERSTRASS and then, with more details, by DARBOUX[7]. Consider a minimal surface S given by the formulas of WEIERSTRASS:

$$\left.\begin{array}{l} x = \Re \int^w (\Phi^2 - \Psi^2)\, dw, \\[2mm] y = \Re \int^w i(\Phi^2 + \Psi^2)\, dw, \\[2mm] z = \Re \int^w 2\, \Phi\Psi\, dw, \end{array}\right\} \tag{5.1}$$

[1] A. HAAR: Über adjungierte Variationsprobleme und adjungierte Extremalflächen. Math. Ann. Vol. 100 (1928) pp. 481—502.

[2] Zum analytischen Charakter usw. Math. Z. Vol. 30 (1929) pp. 404 to 413.

[3] Über Minimalflächen, deren Randkurven wenig von ebenen Kurven abweichen. Abh. preuß. Akad. Wiss. 1909.

[4] Die Lösung des PLATEAUschen Problems über konvexen Bereichen. Math. Ann. Vol. 94 (1925) pp. 53—96.

[5] See, also for references to his previous work, Sur l'intégration des équations aux dérivées partielles du type elliptique. Math. Ann. Vol. 96 (1927) pp. 633—647.

[6] Sur le problème de PLATEAU. Ann. École norm. Vol. 45 (1928) pp. 53—144.

[7] Théorie générale des surfaces Vol. 1 Chapter 13.

and suppose that $w = u + iv$ varies in the upper half-plane $v > 0$. Reflect S upon some plane $\mathfrak{p}$. The resulting minimal surface S^* corresponds then to a couple of new functions Φ^*, Ψ^*, and we have the simple and fundamental theorem that $\Phi^* = a\Phi + b\Psi$, $\Psi^* = c\Phi + d\Psi$, where a, b, c, d are real constants such that $ad - bc = -1$. [1] These constants a, b, c, d are furthermore univocally determined by the plane $\mathfrak{p}$. Consider next a minimal surface $\tilde{S}$ obtained from S by a rigid motion. Since a rigid motion can be obtained by combining an even number of reflections upon properly chosen planes, it follows that $\tilde{S}$ corresponds to a couple of functions $\tilde{\Phi} = \alpha\Phi + \beta\Psi$, $\tilde{\Psi} = \gamma\Phi + \delta\Psi$, where $\alpha, \beta, \gamma, \delta$ are real constants, univocally determined by the rigid motion, such that $\alpha\delta - \beta\gamma = +1$.

V.2. Suppose now that the minimal surface S, given by the formulas (5.1), is bounded by a polygon $\mathfrak{p}$, with n vertices $P_1, P_2, \ldots, P_n$. Suppose also that the correspondence between S and the half-plane $v > 0$ remains one-to-one and continuous on the boundary. Denote by $w_1, w_2, \ldots, w_n$ the images of $P_1, P_2, \ldots, P_n$. Suppose also that Φ, Ψ remain analytic on the open segments of the u-axis bounded by these images. Reflect S upon the side $P_{k-1}P_k$. The resulting surface S_k^- is then, on account of the principle of symmetry (see II.23), an analytic continuation of S, and it follows that Φ, Ψ admit of an analytic continuation through the segment $w_{k-1}w_k$ into the lower half-plane. Denote by Φ_k^-, Ψ_k^- the resulting functions. The same process, applied to the side P_kP_{k+1}, leads to a surface S_k^+ and to a couple of functions Φ_k^+, Ψ_k^+ in the lower half-plane. Since S_k^- and S_k^+ can be obtained from each other by a rigid motion, we have a linear relation between the couples Φ_k^-, Ψ_k^- and Φ_k^+, Ψ_k^+, which permits us to conclude that if the original couple Φ_k, Ψ_k is continued along a closed curve enclosing w_k, then there results a new couple $\Phi_k = \alpha_k\Phi + \beta_k\Psi$, $\Psi_k = \gamma_k\Phi + \delta_k\Psi$, where the real constants $\alpha_k, \beta_k, \gamma_k, \delta_k$ satisfy $\alpha_k\delta_k - \beta_k\gamma_k = +1$ and are univocally determined by the directions of the sides $P_{k-1}P_k$ and P_kP_{k+1} of the polygon $\mathfrak{p}$.

V.3. Thus the couple Φ, Ψ in the formulas (5.1) undergo, by continuation along closed curves around $w_1, \ldots, w_n$, substitutions which are perfectly determined by the directions of the sides of the polygon $\mathfrak{p}$. Further conditions on Φ, Ψ can be conveniently expressed in terms of the differential equation
$$\lambda'' + p\lambda' + q\lambda = 0$$
with coefficients
$$p = -\frac{\Phi\Psi''' - \Psi\Phi'''}{\Phi\Psi'' - \Psi\Phi''}, \qquad q = \frac{\Phi'\Psi''' - \Psi'\Phi'''}{\Phi\Psi'' - \Psi\Phi''},$$
which admits of the solutions $\lambda = \Phi$, $\lambda = \Psi$. It follows, from the behavior of Φ, Ψ in the vicinity of $w_1, \ldots, w_n$, that p, q are single-

---

[1] Darboux: Théorie générale des surfaces Vol. 1 Chapter 13.

valued in the whole plane and have only a finite number of poles as singularities. Hence p, q are rational functions of w. From the condition that the minimal surface S is bounded by a polygon, it follows further that p, q are both real on the w-axis[1].

V.4. These considerations suggest the following plan of attack. Determine first a differential equation $\lambda'' + p\lambda' + q\lambda = 0$, with rational coefficients $p(w)$, $q(w)$, which admits of a couple of solutions $\Phi(w)$, $\Psi(w)$ which undergo, for continuations around n points w_1, w_2, $\ldots$, w_n given on the u-axis, given substitutions, these substitutions being perfectly determined by the directions of the sides of the given polygon $\mathfrak{p}$. The coefficients p, q must be real on the u-axis. The points w_1, w_2, $\ldots$, w_n are, on the other hand, not univocally determined by the directions of the sides of $\mathfrak{p}$. Determine these points w_1, w_2, $\ldots$, w_n in such a way that the minimal surface S, corresponding to Φ, Ψ by means of the formulas (5.1), be bounded by a polygon the sides of which have also the same lengths as the sides of the given polygon $\mathfrak{p}$. This is the program outlined by WEIERSTRASS and DARBOUX, and carried out by R. GARNIER.

V.5. The first step of this program requires the solution (in somewhat restricted form) of the so-called problem of RIEMANN on differential equations with prescribed group of monodromy[2]. The second step requires a delicate investigation of the behavior of the solution of the problem of RIEMANN in its dependence upon the given points w_1, w_2, $\ldots$, w_n. Thus exactly the fine points of the work of GARNIER are beyond the scope of the present report, and we have to restrict ourselves to a few remarks concerning the generality of his result.

V.6. The solution of the problem of RIEMANN secures the existence of a couple Φ, Ψ such that the formulas (5.1) determine a minimal surface S bounded by a polygon $\mathfrak{p}$ the sides of which have the prescribed directions. The next problem is to determine the points w_1, w_2, $\ldots$, w_n in such a way that the sides of $\mathfrak{p}$ have the prescribed lengths. Suppose this problem is known to be solvable for a certain value of n, the other data of the problem being arbitrary. Take a polygon $\mathfrak{p}_{n+1}$ with $n+1$ vertices P_1, P_2, $\ldots$, P_n, P_{n+1}. Denote by $\mathfrak{p}_n$ the polygon with vertices P_1, P_2, $\ldots$, P_n. Let P_{n+1} move, on the side $P_1 P_{n+1}$, toward P_n into a position P_{n+1}^*. For the polygon $\mathfrak{p}_n$ the problem is solvable by assumption. From this GARNIER infers that the problem is possible for the polygon with vertices P_1, P_2, $\ldots$, P_n, P_{n+1}^* if P_{n+1}^* is sufficiently close to P_n. Next he shows that the solution cannot cease to exist if P_{n+1}^*

---

[1] A beautiful presentation of the theory, on the basis of the older literature, is given in the classical work of DARBOUX: Théorie générale des surfaces Vol. 1 Chapter 13.

[2] See, also for literature, R. GARNIER: Solution du problème de RIEMANN. Ann. École norm. Vol. 43 (1926) pp. 177—307.

moves away from P_n on the side $P_n P_{n+1}$. It follows thus that the solution exists for $\mathfrak{p}_{n+1}$ if it is known to exist for every $\mathfrak{p}_n$. Thus it is sufficient to solve the problem for $n = 4$, in which case GARNIER shows that the existence of the solution can be verified in a direct way.

V.7. The above rough description is sufficient to explain an important fact. Suppose we apply the reasoning of GARNIER to a *knotted* polygon $\mathfrak{p}$. Then we have to deform $\mathfrak{p}$ continuously so as to obtain finally a quadrilateral, and consequently we forcibly pass through a position in which the polygon intersects itself. Well then, for polygons with multiple points problem P_3 is generally impossible. Suppose, for instance, the polygon $\mathfrak{p}$ is in the xy-plane and has the shape of an 8. Then problem P_3 reduces (see III.18) to the problem of determining a function $f(w)$ with the following properties.

1. $f(w)$ is analytic in $|w| < 1$.

2. $f(w)$ is continuous in $|w| \leq 1$, and the equation $x + iy = f(w)$ carries $|w| = 1$ into the polygon $\mathfrak{p}$.

The least restrictive interpretation, consistent with the purposes of GARNIER, of condition 2 would be this: every point of $|w| = 1$ is carried into a point on $\mathfrak{p}$, and every point on $\mathfrak{p}$ is image of at least one point of $|w| = 1$. The function $f(w)$ is then not constant, and hence the image D of $|w| < 1$ is a domain (connected open set), such that $\mathfrak{p}$ is the complete boundary of D. If D_1 and D_2 are the two bounded domains bounded by the 8-shaped polygon $\mathfrak{p}$, then every point of D is in D_1 or D_2, and both D_1 and D_2 actually contain points of D. This situation contradicts the fact that D is a connected open set.

As a consequence, the method of GARNIER *gives the solution of problem* P_3 *only in the case of polygons which are not knotted.*

V.8. GARNIER extends his result from polygons to more general curves by a passage to the limit. Let $\varGamma^*$ be a given JORDAN curve, and let $\mathfrak{p}_n$ be an inscribed polygon with n vertices. Let $\varPhi_n$, $\varPsi_n$ be the functions corresponding, in the method of GARNIER, to the polygon $\mathfrak{p}_n$. Insert a new vertex, denote by $\mathfrak{p}_{n+1}$ the resulting polygon and by $\varPhi_{n+1}$, $\varPsi_{n+1}$ the corresponding functions. Under the assumption that $\varGamma^*$ consists of a finite number of arcs with bounded curvature, GARNIER is able to estimate the deviation of the couple $\varPhi_{n+1}$, $\varPsi_{n+1}$ from the couple $\varPhi_n$, $\varPsi_n$ with sufficient accuracy for the purposes of the passage to the limit. The discussion is based again upon the investigation of the analytic dependence of the solution of the problem of RIEMANN upon the data, as this dependence appears on the basis of the method developed by BIRKHOFF[1]. The fine points of the theory belong to a method which is beyond the scope of this report. As far as the geometrical

---

[1] Trans. Amer. Math. Soc. Vol. 10 (1909) pp. 436—470 and Vol. 12 (1911) pp. 243—284.

result is concerned, the whole discussion could be greatly simplified and generalized by applying the reasoning of V.10. It would follow in this way that *problem P_3 is solvable for every not knotted* JORDAN *curve*. It is not known at present if problem P_3 is or is not solvable for every JORDAN curve.

GARNIER also applies his method to the so-called mixed problem which calls for a minimal surface S subject to boundary conditions of the following type. The boundary Γ^* of S consists of a finite number of arcs $a_1, a_2, \ldots; b_1, b_2, \ldots$ The arcs a are given straight segments. The arcs b are not given, but they are required to be situated in given planes, and S is required to be perpendicular to these planes. The geometrical aspects of this problem are described in the classical work of DARBOUX[1]. The method of GARNIER is again based on an investigation of the solution of the problem of RIEMANN which cannot possibly be discussed in this report.

V.9. Problem P_2 (see III.5) calls for three functions $x(u, v), y(u, v), z(u, v)$ with the following properties.

1. $x(u, v), y(u, v), z(u, v)$ are harmonic in $u^2 + v^2 < 1$ and
2. satisfy there the relations $E = G$, $F = 0$, where
$$E = x_u^2 + y_u^2 + z_u^2, \quad F = x_u x_v + y_u y_v + z_u z_v, \quad G = x_v^2 + y_v^2 + z_v^2.$$

3. $x(u, v), y(u, v), z(u, v)$ are continuous in $u^2 + v^2 \leq 1$, and the equations $x = x(u, v), y = y(u, v), z = z(u, v)$ carry $u^2 + v^2 = 1$ in a one-to-one and continuous way into a given JORDAN curve Γ^*.

The methods developed for the solution of this problem work only under various additional assumptions concerning Γ^*. The following theorem is therefore of great importance.

Approximation theorem. *If a sequence Γ_n^* of* JORDAN *curves converges, in the sense of* FRÉCHET, *to a* JORDAN *curve Γ^*, and if problem P_2 is solvable for every curve of the sequence, then the problem is solvable also for the limit curve Γ^*.*

In this generality, the theorem has been proved by J. DOUGLAS[2]. T. RADÓ[3] proved the theorem under the assumption that the lengths of the curves Γ_n^* are uniformly bounded.

A JORDAN curve obviously is the limit, in the sense of FRÉCHET, of a sequence of simple closed polygons. *Hence, on account of the approximation theorem, it is sufficient to solve problem P_2 for polygons.*

V.10. The proof of the approximation theorem may be sketched as follows. Take three distinct points A, B, C on the unit circle $u^2 + v^2 = 1$, and three distinct points A^*, B^*, C^* on Γ^*. Since $\Gamma_n^* \to \Gamma^*$,

---

[1] Théorie générale des surfaces Vol. 1 Chapter 12.

[2] Solution of the problem of PLATEAU. Trans. Amer. Math. Soc. Vol. 33 (1931) pp. 302—306.

[3] Some remarks on the problem of PLATEAU. Proc. Nat. Acad. Sci. U. S. A. Vol. 16 (1930) pp. 242—248.

it is then possible to take three distinct points A_n^*, B_n^*, C_n^* on Γ_n^* such that $A_n^* \to A^*$, $B_n^* \to B^*$, $C_n^* \to C^*$. This being done, consider a solution $x_n(u, v)$, $y_n(u, v)$, $z_n(u, v)$ of problem P_2 for the curve Γ_n^* (the solution exists by assumption). It can be supposed that the equations

$$x = x_n(u, v), \quad y = y_n(u, v), \quad z = z_n(u, v)$$

carry the points A, B, C into the points A_n^*, B_n^*, C_n^*; we say then that we have a uniformly normalized sequence of solutions. Denote, for greater clarity, by $\xi_n(\Theta)$, $\eta_n(\Theta)$, $\zeta_n(\Theta)$ the boundary values of $x_n(u, v)$, $y_n(u, v)$, $z_n(u, v)$ on the unit circle $u = \cos\Theta$, $v = \sin\Theta$. The equations

$$x = \xi_n(\Theta), \quad y = \eta_n(\Theta), \quad z = \zeta_n(\Theta)$$

define then a one-to-one and continuous transformation T_n of $u^2 + v^2 = 1$ into Γ_n^*. All the assumptions of the selection theorem of I.22 being satisfied, there exists an everywhere convergent subsequence of these transformations. Choose a convergent subsequence and use for it the same notation as for the whole sequence. Denote by $\xi(\Theta)$, $\eta(\Theta)$, $\zeta(\Theta)$ the limits of $\xi_n(\Theta)$, $\eta_n(\Theta)$, $\zeta_n(\Theta)$. Then the equations

$$x = \xi(\Theta), \quad y = \eta(\Theta), \quad z = \zeta(\Theta)$$

define a monotonic transformation (see I.21) of $u^2 + v^2 = 1$ into a set on Γ^*. Let $x(u, v)$, $y(u, v)$, $z(u, v)$ be the harmonic functions obtained by means of the POISSON integral formula, using $\xi(\Theta)$, $\eta(\Theta)$, $\zeta(\Theta)$ as boundary functions. These harmonic functions solve then problem P_2 for the limit curve Γ^*. To see this, observe first that the sequences $\xi_n(\Theta)$, $\eta_n(\Theta)$, $\zeta_n(\Theta)$ are uniformly bounded, so that it is legitimate to make a passage to the limit under the integral sign in the formula of POISSON. Hence $x_n(u, v)$, $y_n(u, v)$, $z_n(u, v)$ and all their partial derivatives converge, in $u^2 + v^2 < 1$, to $x(u, v)$, $y(u, v)$, $z(u, v)$ and their partial derivatives respectively. From the relations

$$E_n = G_n, \quad F_n = 0 \quad \text{in} \quad u^2 + v^2 < 1$$

it follows therefore that

$$E = G, \quad F = 0 \quad \text{in} \quad u^2 + v^2 < 1.$$

It remains to investigate what happens on $u^2 + v^2 = 1$. The first step consists in proving that $\xi(\Theta)$, $\eta(\Theta)$, $\zeta(\Theta)$ are continuous. Since these functions define a monotonic transformation of $u^2 + v^2 = 1$ into Γ^*, they have definite one-sided limits $\xi^+(\Theta)$, $\eta^+(\Theta)$, $\zeta^+(\Theta)$, $\xi^-(\Theta)$, $\eta^-(\Theta)$, $\zeta^-(\Theta)$ for every Θ, and their continuity will be proved if it is shown that

$$\xi^+(\Theta) = \xi^-(\Theta), \quad \eta^+(\Theta) = \eta^-(\Theta), \quad \zeta^+(\Theta) = \zeta^-(\Theta)$$

for every Θ (see I.23). DOUGLAS does this in the following way[1]. Without loss of generality, it can be supposed that $\Theta = 0$. Take a point P on

---

[1] A new proof can be obtained by applying the generalized theorem of LINDE-LÖF, stated in III.28.

the positive u-axis, to the right of $u = 1$. Let t denote the transformation obtained by reflecting first on the circle with center at P and orthogonal to the unit circle, and reflecting second on the u-axis. t is then obviously a linear transformation carrying the unit circle into itself.

t carries therefore $x(u, v)$, $y(u, v)$, $z(u, v)$ into three functions $\bar{x}(u, v)$, $\bar{y}(u, v)$, $\bar{z}(u, v)$, which are harmonic in $u^2 + v^2 < 1$, and satisfy there $\bar{E} = \bar{G}$, $\bar{F} = 0$. Furthermore, if $\xi(\Theta)$, $\eta(\Theta)$, $\zeta(\Theta)$ are carried by t into $\bar{\xi}(\Theta)$, $\bar{\eta}(\Theta)$, $\bar{\zeta}(\Theta)$, then $\bar{x}(u, v)$, $\bar{y}(u, v)$, $\bar{z}(u, v)$ are given by the POISSON integral formula, with $\bar{\xi}(\Theta)$, $\bar{\eta}(\Theta)$, $\bar{\zeta}(\Theta)$ as boundary functions. Let now the point P converge to $u = 1$, and watch $\bar{\xi}(\Theta)$ for instance. Obviously, $\bar{\xi}(\Theta)$ converges to $\xi^-(0)$ on the upper half of $u^2 + v^2 = 1$, and to $\xi^+(0)$ on the lower half of $u^2 + v^2 = 1$. Consequently, $\bar{x}(u, v)$ converges in $u^2 + v^2 < 1$ to the harmonic function $x_0(u, v)$ which is obtained from the POISSON integral formula by using the boundary function $\xi_0(\Theta)$ equal to $\xi^-(0)$ for $0 < \Theta < \pi$, and equal to $\xi^+(0)$ for $\pi < \Theta < 2\pi$. The integral can then easily be computed explicitly; it follows that

$$x_0(u, v) = \Re\left(-i\frac{\xi^-(0) - \xi^+(0)}{\pi}\log\frac{1 + w}{1 - w} + \frac{\xi^-(0) - \xi^+(0)}{2}\right),$$

and similar expressions are obtained for the harmonic functions $y_0(u, v)$, $z_0(u, v)$ which are the limits of $\bar{y}(u, v)$, $\bar{z}(u, v)$. As $\bar{x}(u, v)$, $\bar{y}(u, v)$, $\bar{z}(u, v)$ satisfy, in $u^2 + v^2 < 1$, the relations $\bar{E} = \bar{G}$, $\bar{F} = 0$, the limit functions $x_0(u, v)$, $y_0(u, v)$, $z_0(u, v)$ must satisfy the relations $E_0 = G_0$, $F_0 = 0$. From the explicit expressions for $x_0(u, v)$, $y_0(u, v)$, $z_0(u, v)$, it follows by actual computation that

$$E_0 - G_0 - 2iF_0 = -\frac{4}{\pi^2(1 - w^2)^2}[(\xi^-(0) - \xi^+(0))^2 \\ + (\eta^-(0) - \eta^+(0))^2 + (\zeta^-(0) - \zeta^+(0))^2].$$

Consequently, the bracket on the right-hand side vanishes. Hence, $\xi^-(0) = \xi^+(0)$, $\eta^-(0) = \eta^+(0)$, $\zeta^-(0) = \zeta^+(0)$.

The continuity of the boundary functions $\xi(\Theta)$, $\eta(\Theta)$, $\zeta(\Theta)$ being thus established, it follows that the corresponding harmonic functions remain continuous in the closed unit circle and that

$$x(u, v) = \xi(\Theta), \quad y(u, v) = \eta(\Theta), \quad z(u, v) = \zeta(\Theta)$$

on the perimeter of the unit circle, on account of well-known theorems concerning the POISSON integral formula.

It remains to be shown that the equations

$$x = \xi(\Theta), \quad y = \eta(\Theta), \quad z = \zeta(\Theta)$$

carry distinct points of $u^2 + v^2 = 1$ into distinct points of Γ^*. If this were not true, then it would follow (see I.23) that there exists an arc σ

on $u^2 + v^2 = 1$, such that $\xi(\Theta)$, $\eta(\Theta)$, $\zeta(\Theta)$ reduce all three to constants on σ. Without loss of generality, it can be supposed that $\xi(\Theta) = 0$, $\eta(\Theta) = 0$, $\zeta(\Theta) = 0$ on σ. Then[1] (principle of symmetry) the harmonic functions $x(u, v)$, $y(u, v)$, $z(u, v)$ remain analytic on σ, and consequently the relations $E = G$, $F = 0$ hold on σ also. Since $x(u, v) = 0$, $y(u, v) = 0$, $z(u, v) = 0$ on σ, it follows that

$$\frac{\partial x}{\partial \Theta} = -x_u \sin \Theta + x_v \cos \Theta = 0,$$

$$\frac{\partial y}{\partial \Theta} = -y_u \sin \Theta + y_v \cos \Theta = 0,$$

$$\frac{\partial z}{\partial \Theta} = -z_u \sin \Theta + z_v \cos \Theta = 0$$

on σ. Squaring and adding, we obtain $E = G = 0$ on σ, and consequently $x_u = x_v = y_u = y_v = z_u = z_v = 0$ on σ. That is to say, the functions $x_u - i x_v$, $y_u - i y_v$, $z_u - i z_v$, which are analytic functions of $w = u + iv$, vanish on a whole arc in their domain of regularity; consequently, they vanish identically. Hence, $x(u, v)$, $y(u, v)$, $z(u, v)$ and consequently the boundary functions $\xi(\Theta)$, $\eta(\Theta)$, $\zeta(\Theta)$ are constants. This contradicts however the fact that the equations $x = \xi(\Theta)$, $y = \eta(\Theta)$, $z = \zeta(\Theta)$ carry the points A, B, C on $u^2 + v^2 = 1$ into three distinct points A^*, B^*, C^* on Γ^*.

V.11. In dealing with the special case when the lengths of the curves Γ_n^* are uniformly bounded, RADÓ[2] observed that the subsequence $x_n(u, v)$, $y_n(u, v)$, $z_n(u, v)$, used in the proof of the approximation theorem, converges uniformly. According to a remark due to McSHANE (see the end of VI.10), this fact is quite general. Consider, indeed, the convergent subsequence $\xi_n(\Theta)$, $\eta_n(\Theta)$, $\zeta_n(\Theta)$, used in the proof of the approximation theorem, and denote by T_n the monotonic transformation defined by these functions. It has been proved above that the limit functions $\xi(\Theta)$, $\eta(\Theta)$, $\zeta(\Theta)$ define a continuous monotonic transformation.

Now, for sequences of monotonic functions we have the theorem that if such a sequence converges to a continuous function on a closed interval, then the convergence is uniform[3], and the proof of this theorem remains valid for sequences of monotonic transformations. Hence, $\xi_n(\Theta)$, for instance, converges uniformly to $\xi(\Theta)$, and on account of the principle of the maximum it follows that $x_n(u, v)$ converges uniformly to $x(u, v)$ in $u^2 + v^2 \leq 1$. The approximation theorem may

---

[1] The following proof, taken from T. RADÓ: Some remarks on the problem of PLATEAU. Proc. Nat. Acad. Sci. U. S. A Vol. 16 (1930) pp. 242—248, is simpler than that given by DOUGLAS.

[2] Some remarks on the problem of PLATEAU. Proc. Nat. Acad. Sci. U. S. A. Vol. 16 (1930) pp. 242—248.

[3] See for instance PÓLYA-SZEGÖ: Aufgaben und Lehrsätze Vol. 1 p. 63, problem 127.

therefore be completed by the statement that the uniformly normalized sequence of the solutions of problem P_2 for the curves Γ_n^* contains a subsequence which converges *uniformly* to a solution for the limit curve Γ^*. In this form, the result may be considered as a generalization of a theorem, stated by CARATHÉODORY, on the conformal maps of sequences of plane JORDAN regions[1].

If it is known that the solution of problem P_2 for the limit curve Γ^* is unique, then it obviously follows that the whole sequence of the solutions corresponding to the curves Γ_n^* is uniformly convergent. The uniqueness theorem considered in III.11 gives examples for such situations.

V.12. We go on at present to review the methods developed for proving the existence of the solution of problem P_2. We first consider the method due to DOUGLAS[2]. Consider a monotonic transformation (see I.21)

$$T: x = \xi(\Theta), \quad y = \eta(\Theta), \quad z = \zeta(\Theta)$$

of the unit circle $u^2 + v^2 = 1$ into a set on Γ^*. The totality of these transformations constitutes the range of the argument of the functional $A(T)$, introduced by DOUGLAS, and defined by the formula

$$A(T) = \frac{1}{4\pi} \int_0^{2\pi} \int_0^{2\pi} \frac{[\xi(\Theta) - \xi(\psi)]^2 + [\eta(\Theta) - \eta(\psi)]^2 + [\zeta(\Theta) - \zeta(\psi)]^2}{4 \sin^2 \dfrac{\Theta - \psi}{2}} \, d\Theta \, d\psi.$$

The geometrical meaning of the integrand is as follows. Denote by l the length of the secant of the unit circle which is bounded by the points corresponding to the polar angles Θ and ψ, and denote by l^* the length of the corresponding secant of the JORDAN curve Γ^*. Then the integrand is $\left(\dfrac{l^*}{l}\right)^2$. This permits to show that if Γ^* is rectifiable, then there are admissible transformations T for which $A(T)$ is finite. Indeed, in this case we have a one-to-one and continuous correspondence T_0 between $u^2 + v^2 = 1$ and Γ^* such that the ratio of corresponding arcs is equal to $\dfrac{L}{2\pi}$, where L is the length of Γ^*. For T_0 the integrand of the functional of DOUGLAS is clearly bounded, and hence $A(T_0)$ is finite.

The method of DOUGLAS applies to JORDAN curves Γ^* such that there are admissible transformations T for which $A(T)$ is finite. The exact geometrical meaning of this assumption will be considered later; for the time being, it is sufficient to know that every rectifiable curve satisfies this condition. On account of the approximation theorem it

[1] See R. COURANT: Über eine Eigenschaft der Abbildungsfunktionen bei konformer Abbildung. Nachr. Ges. Wiss. Göttingen 1914 and 1922. — T. RADÓ: Sur la représentation conforme de domaines variables. Acta Litt. Sci. Szeged Vol. 1 (1923) pp. 180—186.

[2] Solution of the problem of PLATEAU. Trans. Amer. Math. Soc. Vol. 33 (1931) pp. 263—321.

would be sufficient to know that the method applies to every simple closed polygon.

V.13. Several important properties of the functional $A(T)$ become obvious when we consider another expression of $A(T)$. Using the FOURIER series of $\xi(\Theta)$, $\eta(\Theta)$, $\zeta(\Theta)$, DOUGLAS finds that $A(T)$ admits of the following equivalent definition. Let $x(u, v)$, $y(u, v)$, $z(u, v)$ denote the harmonic functions corresponding to $\xi(\Theta)$, $\eta(\Theta)$, $\zeta(\Theta)$ by means of the POISSON integral formula. Then

$$\left. \begin{aligned} A(T) &= \tfrac{1}{2} \iint_{u^2+v^2<1} (E+G)\, du\, dv \\ &= \tfrac{1}{2}\left[\iint_{u^2+v^2<1} (x_u^2+x_v^2)\, du\, dv + \iint_{u^2+v^2<1} (y_u^2+y_v^2)\, du\, dv + \iint_{u^2+v^2<1} (z_u^2+z_v^2)\, du\, dv \right] \end{aligned} \right\}. \quad (5.2)$$

In other words: $A(T)$ is half the sum of the DIRICHLET integrals of the harmonic functions $x(u, v)$, $y(u, v)$, $z(u, v)$.

V.14. Invariance of $A(T)$.[1] Let

$$T: x = \xi(\Theta), \quad y = \eta(\Theta), \quad z = \zeta(\Theta), \quad\quad (5.3)$$

be a monotonic transformation of $u^2 + v^2 = 1$ into a set on Γ^*. Map the unit circle $u^2 + v^2 \leq 1$ upon itself in a one-to-one and conformal way. The functions $\xi(\Theta)$, $\eta(\Theta)$, $\zeta(\Theta)$ are carried into new functions $\bar{\xi}(\Theta)$, $\bar{\eta}(\Theta)$, $\bar{\zeta}(\Theta)$, and

$$\overline{T}: x = \overline{\xi}(\Theta), \quad y = \bar{\eta}(\Theta), \quad z = \overline{\zeta}(\Theta)$$

is again a monotonic transformation. Two transformations T, $\overline{T}$ related in this manner are called equivalent by DOUGLAS.

If T and $\overline{T}$ are equivalent, then $A(T) = A(\overline{T})$. This is obvious from the expression (5.2) of $A(T)$, since the DIRICHLET integral is invariant under conformal mapping.

V.15. Effect of discontinuities of T on $A(T)$. Let T be given by (5.3). If $\xi(\Theta)$, for instance, is discontinuous at a certain point Θ_0, then with regard to I.23 this can happen only in the following manner: $\xi(\Theta)$ has definite one-sided limits ξ_0^+, ξ_0^- at Θ_0, and these limits are different.

In his work on conformal mapping, COURANT[2] used a simple and important reasoning which leads to the following lemma[3].

Let $\xi(\Theta)$ be integrable (in the RIEMANN sense, for instance), and let $x(u, v)$ be defined by the POISSON integral:

$$x(u, v) = \frac{1}{2\pi}\int_0^{2\pi} \xi(\Theta)\, \frac{1-r^2}{1+r^2-2r\cos(\Theta-\varphi)}\, d\Theta, \quad u = r\cos\varphi, \quad v = r\sin\varphi.$$

---

[1] The reasoning in V.14 to V.16 is somewhat simpler than that used by DOUGLAS.

[2] See HURWITZ-COURANT: Funktionentheorie, pp. 375—376.

[3] See T. RADÓ: The problem of the least area and the problem of PLATEAU. Math. Z. Vol. 32 (1930) pp. 776—779.

Suppose $\xi(\Theta)$ has definite one-sided limits ξ_0^+, ξ_0^- for some value Θ_0 of Θ, such that $\xi_0^+ \neq \xi_0^-$. Then the DIRICHLET integral

$$\iint_{u^2 + v^2 < 1} (x_u^2 + x_v^2)\, du\, dv$$

cannot be finite.

Applying this to the function $\xi(\Theta)$ figuring in the definition of the transformation T, we obtain from the expression (5.2) of $A(T)$ that if $\xi(\Theta)$ has a discontinuity, then $A(T)$ cannot be finite. The same remark applies of course to $\eta(\Theta)$, $\zeta(\Theta)$. Hence:

If $A(T)$ is finite, then the transformation T necessarily is continuous.

V.16. *Lower semi-continuity of $A(T)$.* Consider a sequence

$$T_n: x = \xi_n(\Theta), \quad y = \eta_n(\Theta), \quad z = \zeta_n(\Theta)$$

of monotonic transformations. Suppose T_n converges to a monotonic transformation

$$T: x = \xi(\Theta), \ y = \eta(\Theta), \ z = \zeta(\Theta).$$

Then
$$A(T) \leqq \underline{\lim} A(T_n). \tag{5.4}$$

Denote by $x_n(u, v)$, $y_n(u, v)$, $z_n(u, v)$ the harmonic functions corresponding to $\xi_n(\Theta)$, $\eta_n(\Theta)$, $\zeta_n(\Theta)$ by means of the POISSON integral formula. Let $x(u, v)$, $y(u, v)$, $z(u, v)$ have the same meaning with respect to $\xi(\Theta)$, $\eta(\Theta)$, $\zeta(\Theta)$. Then it follows from the POISSON integral formula that $x_n(u, v)$, $y_n(u, v)$, $z_n(u, v)$ and their partial derivatives converge to $x(u, v)$, $y(u, v)$, $z(u, v)$ and their partial derivatives respectively in $u^2 + v^2 < 1$, and that the convergence is uniform in every concentric circle $u^2 + v^2 \leqq r^2 < 1$. Hence, for $0 < r < 1$:

$$\tfrac{1}{2}\iint_{u^2+v^2<r^2} (E + G)\, du\, dv = \lim \tfrac{1}{2}\iint_{u^2+v^2<r^2} (E_n + G_n)\, du\, dv \leqq \lim \tfrac{1}{2}\iint_{u^2+v^2<1} (E_n + G_n)\, du\, dv.$$

The passage to the limit $r \to 1$ yields the inequality (5.4).

V.17. The existence proof of DOUGLAS proceeds then in the following steps. Let m denote the (by assumption finite) greatest lower bound of the functional $A(T)$. There exists a sequence $\overline{T}_n$ such that $A(\overline{T}_n) \to m$. Of course, only such transformations $\overline{T}_n$ will be used for which $A(\overline{T}_n)$ is finite. Hence, $\overline{T}_n$ is continuous (see V.15). Choose now three distinct points A, B, C on $u^2 + v^2 = 1$, and three distinct fixed points A^*, B^*, C^* on the given JORDAN curve Γ^*. Since $\overline{T}_n$ is continuous, A^*, B^*, C^* will be the images, under $\overline{T}_n$, of three distinct points A_n, B_n, C_n. Using a linear transformation of the unit circle into itself which carries A_n, B_n, C_n into A, B, C, we obtain a monotonic transformation T_n which carries A, B, C into A^*, B^*, C^* and which is equivalent to $\overline{T}_n$. Hence $A(T_n) = A(\overline{T}_n)$ (see V.14). In this way we obtain a sequence T_n of monotonic transformations such that $A(T_n) \to m$, and such that every T_n carries the three prescribed points A, B, C of $u^2 + v^2 = 1$

into the three prescribed points A^*, B^*, C^* of Γ^*. The sequence T_n satisfies the assumptions of the selection theorem in I.22, and therefore contains an everywhere convergent subsequence, which we denote by T_n again for simplicity. There results a limit transformation

$$T_0 : x = \xi_0(\Theta), \quad y = \eta_0(\Theta), \quad z = \zeta_0(\Theta), \tag{5.5}$$

which is again a monotonic transformation of $u^2 + v^2 = 1$ into a set on Γ^*. Obviously, T_0 carries the points A, B, C into the points A^*, B^*, C^* respectively. By the definition of m we have $A(T_0) \geq m$, while on the other hand, on account of the lower semi-continuity of the functional of DOUGLAS, $A(T_0) \leq \lim A(T_n) = m$. Hence, $A(T_0) = m$. Since m is finite, T_0 is continuous (see V.15). Summing up:

There exists a transformation T_0 which 1. minimizes the functional $A(T)$, which 2. is continuous, and which 3. carries the three prescribed points A, B, C on $u^2 + v^2 = 1$ into three prescribed points A^*, B^*, C^* on Γ^*.

V.18. Consider this minimizing T_0, given by (5.5). Denote by $x_0(u, v)$, $y_0(u, v)$, $z_0(u, v)$ the harmonic functions which reduce to $\xi_0(\Theta)$, $\eta_0(\Theta)$, $\zeta_0(\Theta)$ on $u^2 + v^2 = 1$. x_0, y_0, z_0 are the real parts of functions Φ_1, Φ_2, Φ_3, which are analytic functions of $w = u + iv$ in $u^2 + v^2 < 1$. Then we have (cf. II.18)

$$E_0 - G_0 - 2iF_0 = \Phi_1'^2 + \Phi_2'^2 + \Phi_3'^2. \tag{5.6}$$

The next step is to show that the minimizing property of T_0 has the consequence that $\Phi_1'^2 + \Phi_2'^2 + \Phi_3'^2 = 0$.

This comes out as follows[1]. By a series of computations, the following formulas are established.

$$\Phi_1'^2 + \Phi_2'^2 + \Phi_3'^2 = \frac{1}{2\pi^2} \int_0^{2\pi} \int_0^{2\pi} \Psi(\varphi, \psi) \frac{dz\, d\zeta}{(z-w)^2(\zeta-w)^2}, \tag{5.7}$$

$$A(T_0) = \frac{1}{4\pi} \int_0^{2\pi} \int_0^{2\pi} \Psi(\varphi, \psi) \frac{dz\, d\zeta}{(z-\zeta)^2},$$

where $z = e^{i\varphi}$, $\zeta = e^{i\psi}$, and

$$\Psi(\varphi, \psi) = [\xi_0(\varphi) - \xi_0(\psi)]^2 + [\eta_0(\varphi) - \eta_0(\psi)]^2 + [\zeta_0(\varphi) - \zeta_0(\psi)]^2.$$

Let then λ be a real parameter and consider the following (for small values of λ obviously one-to-one and continuous) transformations of the unit circle $u = \cos\Theta$, $v = \sin\Theta$ into itself:

$$z' = z \exp \lambda \left| \frac{1}{z(z-w)} - \frac{1}{z(\bar{z}-w)} \right|,$$

and

$$z' = z \exp \lambda \left[\frac{1}{z(z-w)} + \frac{1}{z(\bar{z}-w)} \right],$$

[1] We follow the simplified presentation given by DOUGLAS in a later paper: The problem of PLATEAU for two contours. J. Math. Physics, Massachusetts Inst. Technol. Vol. 10 (1931) pp. 315—359.

where the bar denotes the conjugate complex number, and w is a fixed interior point of the unit circle. Using one of these transformations, we transform $\xi_0(\Theta)$, $\eta_0(\Theta)$, $\zeta_0(\Theta)$ into three new functions $\xi(\Theta)$, $\eta(\Theta)$, $\zeta(\Theta)$; and

$$T : x = \xi(\Theta),\ y = \eta(\Theta),\ z = \zeta(\Theta)$$

is a monotonic transformation which depends upon λ. Consequently $A(T)$ is a function $J(\lambda)$ of λ which has a minimum for $\lambda = 0$. Hence $J'(0) = 0$.

Computation shows that

$$J'(0) = -\frac{1}{4\pi}\int_0^{2\pi}\int_0^{2\pi} \Psi(\varphi, \psi)\, \frac{dz\, d\zeta}{(z - w)^2 (\zeta - w)^2}.$$

Comparing with (5.7), we see that $J'(0) = 0$ implies that $\Phi_1^2 + \Phi_2^2 + \Phi_3^2 = 0$. Consequently, on account of (5.6), $E_0 = G_0$, $F_0 = 0$.

Thus, the harmonic functions $x_0(u, v)$, $y_0(u, v)$, $z_0(u, v)$ satisfy the relations $E_0 = G_0$, $F_0 = 0$ for $u^2 + v^2 < 1$. They are continuous in $u^2 + v^2 \leq 1$. On $u^2 + v^2 = 1$ they reduce to the functions $\xi(\Theta)$, $\eta(\Theta)$, $\zeta(\Theta)$, and consequently the equations

$$x = x_0(u, v),\quad y = y_0(u, v),\quad z = z_0(u, v)$$

carry $u^2 + v^2 = 1$ in a continuous and monotonic way into Γ^*, the points A, B, C of $u^2 + v^2 = 1$ being taken into the prescribed points A^*, B^*, C^* of Γ^*. It remains to be shown that distinct points of $u^2 + v^2 = 1$ are taken into distinct points of Γ^*. If this were not the case, there would exist an arc of $u^2 + v^2 = 1$ on which $x_0(u, v)$, $y_0(u, v)$, $z_0(u, v)$ would all three reduce to constants. Since $E_0 = G_0$, $F_0 = 0$, it would then follow (cf. the end of V.10) that $x_0(u, v)$, $y_0(u, v)$, $z_0(u, v)$ reduce to constants identically. This contradicts however the fact that the points A, B, C are carried into three distinct points A^*, B^*, C^*.

Thus the functions $x_0(u, v)$, $y_0(u, v)$, $z_0(u, v)$ solve problem P_2 for the given JORDAN curve Γ^*. The generality of Γ^* is restricted by the assumption that the greatest lower bound of the functional $A(T)$ is finite. This assumption is however certainly satisfied for every rectifiable JORDAN curve (see V.12), and hence, in particular, for every simple closed polygon. On account of the approximation theorem, this is sufficient to establish the existence of the solution of problem P_2 for every JORDAN curve.

V.19. Problem P_2, as stated above, is considered by DOUGLAS as the three-dimensional special case of the following n-dimensional problem. Determine n functions $x_1(u, v)$, ..., $x_n(u, v)$ with the following properties.

1. $x_1(u, v)$, ..., $x_n(u, v)$ are harmonic for $u^2 + v^2 < 1$ and
2. satisfy there the relations $E = G$, $F = 0$, where

$$E = x_{1u}^2 + \cdots + x_{nu}^2, \quad F = x_{1u}x_{1v} + \cdots + x_{nu}x_{nv}, \quad G = x_{1v}^2 + \cdots + x_{nv}^2.$$

3. $x_1(u, v)$, ..., $x_n(u, v)$ are continuous in $u^2 + v^2 \leqq 1$ and the equations $x_1 = x_1(u, v)$, ..., $x_n = x_n(u, v)$ carry $u^2 + v^2 = 1$ in a one-to-one and continuous way into a JORDAN curve $\varGamma^*$ given in the n-dimensional space.

DOUGLAS carries out his method for a general n and emphasizes the fact that the value of n does not make any difference. For this reason we restricted ourselves, in the preceding review of his method, to the familiar three-dimensional case. Besides the case $n = 3$, the case $n = 2$ is important. Let us first recall the classical *theorem of* DARBOUX[1]. Let R be a closed region in the $w = u + iv$ plane, bounded by a JORDAN curve. Let there be given a function $f(w)$, which is continuous in R and analytic in the interior of R. Suppose that $f(w)$ takes on distinct values at distinct boundary points of R, that is to say suppose that the equation $\zeta = f(w)$ carries the boundary curve of R in a one-to-one and continuous way into a JORDAN curve in the ζ-plane. Denote by R^* the region bounded by this image curve. *The equation $\zeta = f(w)$ carries then R in a one-to-one and continuous way into R^*, and this transformation is conformal in the interior of the regions.*

Consider now the case $n = 2$ of problem P_2. We have then two harmonic functions $x_1(u, v)$, $x_2(u, v)$, and condition 2 of the problem implies that $x_1(u, v)$, $x_2(u, v)$ are conjugate harmonic functions, that is to say real and imaginary part respectively of an analytic function $f(w)$ of $w = u + iv$. On account of the theorem of DARBOUX, it follows that in the case $n = 2$, problem P_2 reduces to the problem of finding a function $\zeta = f(w)$, which maps the unit circle $|w| \leqq 1$ upon a given JORDAN region in a one-to-one, continuous and, in the interior of the regions, conformal way. DOUGLAS points out that his method is entirely independent of the theory of conformal mapping and that consequently his method yields a new solution of this important problem. He also points out that while the modern methods of dealing with the problem establish first the existence of the mapping function for the interior and obtain information as to its behavior on the boundary afterwards, his method leads directly to the correspondence of the boundaries. The behavior of the mapping function in the interior follows then immediately from the theorem of DARBOUX.

Thus the fact that the method of DOUGLAS is independent of the theory of conformal mapping appears as one of its essential features.

V.20. The method of T. RADO[2], which will be reviewed presently, depends essentially on the theory of conformal mapping. The idea of the method is to construct first an *approximate solution of the problem*, and to obtain the exact solution by a *passage to the limit*. The con-

---

[1] See for instance OSGOOD: Lehrbuch der Funktionentheorie Vol. 1, 5. edition pp. 397−399.

[2] On PLATEAU's problem. Ann. of Math. Vol. 31 (1930) pp. 457−469.

struction of the approximate solution being the essential part of the method, it may be interesting to point out the simple facts on which the construction is based. Let $a(\Gamma^*)$ denote the greatest lower bound of the areas of all the surfaces (of the type of the circle) bounded by a given JORDAN curve Γ^*. Given then a $\sigma > 0$, there exists a surface

$$\overline{S} : x = \overline{x}(u, v), \quad y = \overline{y}(u, v), \quad z = \overline{z}(u, v), \quad u^2 + v^2 \leq 1 \qquad (5.8)$$

bounded by Γ^* and such that

$$a(\Gamma^*) \leq \mathfrak{A}(\overline{S}) \leq a(\Gamma^*) + \sigma,$$

where $\mathfrak{A}(\overline{S})$ denotes the area of $\overline{S}$. Make now the (generally unjustified) assumption that $\overline{S}$ admits of a conformal map. Suppose that (5.8) is such a map; then $\overline{E} = \overline{G}$, $\overline{F} = 0$, and the formula

$$\mathfrak{A}(\overline{S}) = \int\int (\overline{E}\overline{G} - \overline{F}^2)^{\frac{1}{2}}$$

reduces to

$$\mathfrak{A}(\overline{S}) = \int\int \overline{E} = \int\int \overline{G} = \tfrac{1}{2}\int\int (\overline{E} + \overline{G}),$$

where the integrals are taken in $u^2 + v^2 < 1$. Denote then by $x(u, v)$, $y(u, v)$, $z(u, v)$ the harmonic functions which coincide, for $u^2 + v^2 = 1$, with $\overline{x}(u, v)$, $\overline{y}(u, v)$, $\overline{z}(u, v)$ respectively. Then

$$S : x = x(u, v), \quad y = y(u, v), \quad z = z(u, v), \quad u^2 + v^2 \leq 1$$

is again bounded by Γ^* and consequently $a(\Gamma^*) \leq \mathfrak{A}(S)$. Since a harmonic function, with prescribed boundary values, minimizes the DIRICHLET integral, we have

$$\left. \begin{aligned} \int\int (x_u^2 + x_v^2) &\leq \int\int (\overline{x}_u^2 + \overline{x}_v^2), \\ \int\int (y_u^2 + y_v^2) &\leq \int\int (\overline{y}_u^2 + \overline{y}_v^2), \\ \int\int (z_u^2 + z_v^2) &\leq \int\int (\overline{z}_u^2 + \overline{z}_v^2). \end{aligned} \right\} \qquad (5.9)$$

Addition gives

$$\int\int (E + G) \leq \int\int (\overline{E} + \overline{G}).$$

Combining all the preceding information, we can write

$$\left. \begin{aligned} a(\Gamma^*) \leq \mathfrak{A}(S) = \int\int (EG - F^2)^{\frac{1}{2}} \leq \int\int E^{\frac{1}{2}} G^{\frac{1}{2}} \leq \tfrac{1}{2}\int\int (E + G) \\ \leq \tfrac{1}{2}\int\int (\overline{E} + \overline{G}) = \mathfrak{A}(\overline{S}) \leq a(\Gamma^*) + \sigma. \end{aligned} \right\} \qquad (5.10)$$

Hence any two of the integrals appearing in this sequence of inequalities differ from each other by not more than σ, and this leads immediately to the inequalities

$$\int\int (E^{\frac{1}{2}} - G^{\frac{1}{2}})^2 \leq 2\sigma,$$

$$\int\int |F| \leq [2\sigma(\sigma + a(\Gamma^*))]^{\frac{1}{2}}.$$

Thus the functions $x(u, v)$, $y(u, v)$, $z(u, v)$ satisfy the conditions of problem P_2, except for the conditions $E = G$, $F = 0$. These conditions are satisfied *approximately* in the sense that the integrals

$$\iint (E^{\frac{1}{2}} - G^{\frac{1}{2}})^2 \quad \text{and} \quad \iint |F|$$

are as small as we please, since σ was arbitrary.

The case $\sigma = 0$ explains the idea and the origin of this construction[1]. If $\sigma = 0$, then we have the sign of equality all over in (5.10), and consequently also in (5.9). This implies however that $x(u, v) \equiv \bar{x}(u, v)$, $y(u, v) \equiv \bar{y}(u, v)$, $z(u, v) \equiv \bar{z}(u, v)$. That is to say, $\bar{x}(u, v)$, $\bar{y}(u, v)$, $\bar{z}(u, v)$ are harmonic functions. *Hence, if we have a surface $\overline{S}$ which admits of a conformal map and whose area is a minimum, then $\overline{S}$ solves the problem of* PLATEAU. *The above construction shows then that if we have a surface S which admits of a conformal map and whose area is approximately a minimum, then we can derive from it an approximate solution of the problem of* PLATEAU.

This construction of the approximate solution is based on the assumption that the surface S admits of a conformal map. In general, however, a surface does *not* admit of a conformal map. On the other hand, there is a great latitude in the choice of S. Further discussion will show that $\overline{S}$ can be replaced, as far as the successful application of the construction is concerned, by a nearby *polyhedron*. Thus it is sufficient to know that polyhedrons admit of conformal maps, and this already has been established by H. A. SCHWARZ. We now are going to describe briefly the actual application of the preceding considerations to the solution of problem P_2.

V.21. Suppose the given JORDAN curve is a simple closed polygon and denote this polygon by $\mathfrak{p}^*$. Consider all the polyhedrons bounded by $\mathfrak{p}^*$ (see I.9) and denote by $\mu(\mathfrak{p}^*)$ the greatest lower bound of their areas. $\mu(\mathfrak{p}^*)$ obviously is finite.

Consider also all the continuous surfaces bounded by $\mathfrak{p}^*$ and denote by $\mathfrak{a}(\mathfrak{p}^*)$ the greatest lower bound of their areas. Obviously, $\mathfrak{a}(\mathfrak{p}^*) \leq \mu(\mathfrak{p}^*)$. In his Thesis[2], LEBESGUE proved a theorem from which it follows that $\mathfrak{a}(\mathfrak{p}^*) = \mu(\mathfrak{p}^*)$. This means that for the area $\mathfrak{A}(S)$ of every continuous surface S, bounded by $\mathfrak{p}^*$, we have $\mathfrak{A}(S) \geq \mu(\mathfrak{p}^*)$. The result of LEBESGUE, referred to above, escaped the attention of T. RADÓ. He makes use of the fact, easily established, that the inequality $\mathfrak{A}(S) \geq \mu(\mathfrak{p}^*)$ holds for harmonic surfaces bounded by $\mathfrak{p}^*$, that is to say for surfaces which admit of a representation $\mathfrak{x} = \mathfrak{x}(u, v)$, $u^2 + v^2 \leq 1$, where the components $x(u, v)$, $y(u, v)$, $z(u, v)$ of $\mathfrak{x}(u, v)$ are harmonic functions.

---

[1] See II.10.

[2] Intégrale, longueur, aire. Ann. Mat. pura appl. Vol. 7 (1902) pp. 231—359.

V.22. The first step is then to solve the following approximate problem. Give three distinct points A, B, C on $u^2 + v^2 = 1$, three distinct points A^*, B^*, C^* on $\mathfrak{p}^*$, and also give an $\varepsilon > 0$. Determine three functions $x(u, v)$, $y(u, v)$, $z(u, v)$ with the following properties.

1. $x(u, v)$, $y(u, v)$, $z(u, v)$ are harmonic for $u^2 + v^2 < 1$, and
2. satisfy the relations

$$\iint (E^{\frac{1}{2}} - G^{\frac{1}{2}})^2 \leqq \varepsilon, \quad \iint |F| \leqq \varepsilon,$$

where the integrals are extended over $u^2 + v^2 < 1$.

3. $x(u, v)$, $y(u, v)$, $z(u, v)$ are continuous in $u^2 + v^2 \leqq 1$, and the equations $x = x(u, v)$, $y = y(u, v)$, $z = z(u, v)$ carry $u^2 + v^2 = 1$ in a one-to-one and continuous way into $\mathfrak{p}^*$, the points A, B, C being carried into the points A^*, B^*, C^*.

This problem is solved as follows. Let $\sigma > 0$ be a constant to be determined later on. By the definition of $\mu(\mathfrak{p}^*)$ (see V.21) there is a polyhedron $\overline{\mathfrak{P}}$, bounded by $\mathfrak{p}^*$, such that $\mathfrak{A}(\overline{\mathfrak{P}}) \leqq \mu(\mathfrak{p}^*) + \sigma$. Let

$$\overline{\mathfrak{P}} : x = \bar{x}(u, v), \quad y = \bar{y}(u, v), \quad z = \bar{z}(u, v), \quad u^2 + v^2 \leqq 1$$

be an isothermic representation of $\overline{\mathfrak{P}}$, such that A, B, C are carried into A^*, B^*, C^*. We have then, on account of $\bar{E} = \bar{G}, \bar{F} = 0$,

$$\mathfrak{A}(\overline{\mathfrak{P}}) = \tfrac{1}{2} \iint (\bar{E} + \bar{G}).$$

Let $x(u, v)$, $y(u, v)$, $z(u, v)$ be the harmonic functions which coincide with $\bar{x}(u, v)$, $\bar{y}(u, v)$, $\bar{z}(u, v)$ respectively on $u^2 + v^2 = 1$. *Then these harmonic functions solve the approximate problem.*

Indeed, conditions 1 and 3 obviously are satisfied. To verify that condition 2 is satisfied, use again the fact that a harmonic function with given boundary values minimizes the DIRICHLET integral. This gives (cf. V.20)

$$\iint (E + G) \leqq \iint (\bar{E} + \bar{G}).$$

The surface

$$S : x = x(u, v), \quad y = y(u, v), \quad z = z(u, v), \quad u^2 + v^2 \leqq 1$$

is a harmonic surface bounded by $\mathfrak{p}^*$. Hence (see V.21)

$$\mu(\mathfrak{p}^*) \leqq \mathfrak{A}(S) = \iint (EG - F^2)^{\frac{1}{2}}.$$

Combining all the preceding information, we can write

$$\left. \begin{aligned} \mu(\mathfrak{p}^*) \leqq \mathfrak{A}(S) &= \iint (EG - F^2)^{\frac{1}{2}} \leqq \iint E^{\frac{1}{2}} G^{\frac{1}{2}} \leqq \tfrac{1}{2} \iint (E + G) \\ &\leqq \tfrac{1}{2} \iint (\bar{E} + \bar{G}) = \mathfrak{A}(\overline{\mathfrak{P}}) \leqq \mu(\mathfrak{p}^*) + \sigma. \end{aligned} \right\}$$

Hence (cf. V.20)

$$\iint (E^{\frac{1}{2}} - G^{\frac{1}{2}})^2 \leqq 2\sigma,$$

$$\iint |F| \leqq [2\sigma(\sigma + \mu(\mathfrak{p}^*))]^{\frac{1}{2}}.$$

By proper choice of σ these bounds can be made less than the prescribed ε.

V.23. The solution of the exact problem P_2 is obtained now by a passage to the limit. Keep the points A, B, C and A^*, B^*, C^* fixed and solve the approximate problem for $\varepsilon_1, \varepsilon_2, \ldots, \varepsilon_n, \ldots$, where $\varepsilon_n \to 0$. There results a sequence of approximate solutions $x_n(u, v)$, $y_n(u, v)$, $z_n(u, v)$ and the problem is to show that a properly chosen subsequence converges to a solution of the exact problem.

Suppose we have obtained, in any manner whatsoever, a subsequence $x_{n_k}(u, v)$, $y_{n_k}(u, v)$, $z_{n_k}(u, v)$ converging in $u^2 + v^2 < 1$ to the (necessarily harmonic) functions $x(u, v)$, $y(u, v)$, $z(u, v)$. The convergence extends then to all partial derivatives, and is uniform in every concentric circle $u^2 + v^2 \leq r^2 < 1$. Hence, r being fixed,

$$\iint\limits_{(r)} (E_n^{\frac{1}{2}} - G_n^{\frac{1}{2}})^2 \to \iint\limits_{(r)} (E^{\frac{1}{2}} - G^{\frac{1}{2}})^2, \quad \iint\limits_{(r)} |F_n| \to \iint\limits_{(r)} |F|,$$

where (r) indicates that the integrals are taken over $u^2 + v^2 \leq r^2$. Since

$$\iint\limits_{(r)} (E_n^{\frac{1}{2}} - G_n^{\frac{1}{2}})^2 \leq \iint\limits_{u^2+v^2<1} (E_n^{\frac{1}{2}} - G_n^{\frac{1}{2}})^2 \leq \varepsilon_n \to 0,$$

$$\iint\limits_{(r)} |F_n| \leq \iint\limits_{u^2+v^2<1} |F_n| \leq \varepsilon_n \to 0,$$

it follows that
$$\iint\limits_{(r)} (E^{\frac{1}{2}} - G^{\frac{1}{2}})^2 = 0, \quad \iint\limits_{(r)} |F| = 0.$$

Since E, F, G are continuous, it follows that $E = G$, $F = 0$ for $u^2 + v^2 < r^2$. Since this is true for every $r < 1$, it follows that $E = G$, $F = 0$ in the whole interior of the unit circle.

Using this simple remark, T. RADÓ verifies that the method he used to prove a special case of the approximation theorem (see V.9), namely the special case when the lengths of the approximating curves are uniformly bounded, works without further modification under the present conditions. The reason for this is the fact that the approximate relations

$$\iint (E_n^{\frac{1}{2}} - G_n^{\frac{1}{2}})^2 \leq \varepsilon_n, \quad \iint |F_n| \leq \varepsilon_n$$

are, for the purposes of the passage to the limit, just as efficient as the exact relations $E = G$, $F = 0$ would be.

In this way follows the existence of the solution of problem P_2 for every simple closed polygon. Using the special case of the approximation theorem he has previously proved, T. RADÓ extends then the existence theorem to rectifiable JORDAN curves.

V.24. The reviewer takes the liberty to call the attention of the reader to a very simple and efficient method resulting from a combination of the proof of the approximation theorem of J. DOUGLAS with the notion and the construction, given above, of the approximate solution. Let the JORDAN curves Γ_n^* converge, in the sense of FRÉCHET, to the JORDAN curve Γ^*. Take three distinct points A, B, C on $u^2 + v^2 = 1$, three distinct points A^*, B^*, C^* on Γ^*, and three distinct points A_n^*, B_n^*, C_n^* on Γ_n^* in such a way that $A_n^* \to A^*$, $B_n^* \to B^*$, $C_n^* \to C^*$ (this is possible since $\Gamma_n^* \to \Gamma^*$). Give also a sequence of positive constants $\varepsilon_n \to 0$. Suppose that it is possible to solve, for every curve of the sequence Γ_n^*, the following approximate problem. Determine three functions $x_n(u, v)$, $y_n(u, v)$, $z_n(u, v)$ with the following properties.

1. $x_n(u, v)$, $y_n(u, v)$, $z_n(u, v)$ are harmonic in $u^2 + v^2 < 1$, and
2. satisfy the relations

$$\int\int (E_n^{\frac{1}{2}} - G_n^{\frac{1}{2}})^2 \leq \varepsilon_n, \quad \int\int |F_n| \leq \varepsilon_n.$$

3. $x_n(u, v)$, $y_n(u, v)$, $z_n(u, v)$ are continuous in $u^2 + v^2 \leq 1$, and the equations $x = x_n(u, v)$, $y = y_n(u, v)$, $z = z_n(u, v)$ carry $u^2 + v^2 = 1$ in a one-to-one and continuous way into Γ_n^*, the points A, B, C being carried into A_n^*, B_n^*, C_n^*.

Apply then to this situation the proof of the approximation theorem of J. DOUGLAS, supplemented by the remarks of V.23 concerning the handling of the approximate relations

$$\int\int (E_n^{\frac{1}{2}} - G_n^{\frac{1}{2}})^2 \leq \varepsilon_n, \quad \int\int |F_n| \leq \varepsilon_n.$$

There follows, without any further change in the proof, that there exists a subsequence converging uniformly, in $u^2 + v^2 \leq 1$, to an exact solution of problem P_2 for the limit curve Γ^*.

Start then with a given JORDAN curve Γ^* and approximate it, in the sense of FRÉCHET, by a sequence of simple closed polygons $\mathfrak{p}_n^*$. Use the method of V.22 to solve the approximate problem for $\mathfrak{p}_n^*$. The generalization of the approximation theorem, described above, yields then the solution of the exact problem P_2 for the given JORDAN curve Γ^*.

V.25. One of the important steps in the method of J. DOUGLAS is the discussion of the first variation of his functional. T. RADÓ observed that this discussion can be based on a simple lemma concerning the conformal maps of surfaces[1].

Let $\mathfrak{A}(S)$ denote the area of a (sufficiently regular) surface

$$S : x = x(u, v), \quad y = y(u, v), \quad z = z(u, v), \quad u^2 + v^2 \leq 1. \quad (5.11)$$

Then $$\mathfrak{A}(S) = \int\int (EG - F^2)^{\frac{1}{2}} \leq \int\int E^{\frac{1}{2}} G^{\frac{1}{2}} \leq \frac{1}{2} \int\int (E + G). \quad (5.12)$$

Hence $$\mathfrak{A}(S) \leq \frac{1}{2} \int\int (E + G)$$

for every representation of S, while for isothermic representations obviously

$$\mathfrak{A}(S) = \frac{1}{2} \int\int (E + G),$$

on account of $E = G, F = 0$. While hasty inferences from these remarks are invalidated by the fact that a surface, in general, does not admit

---

[1] T. RADÓ: On the functional of Mr. DOUGLAS. **Ann. of Math.** Vol. 32 (1931) pp. 785—803.

of an isothermic representation (see I.18), still it is possible to draw a conclusion sufficiently strong for the purposes of the problem of PLATEAU.

Suppose a surface S admits of a representation (5.11) where $x(u, v)$, $v(u, v)$, $z(u, v)$ are continuous for $u^2 + v^2 \leqq 1$ and have continuou-partial derivatives of the first and second order for $u^2 + v^2 < 1$. Ins troduce new parameters $\bar{u}, \bar{v}$ by equations $u = u(\bar{u}, \bar{v})$, $v = v(\bar{u}, \bar{v})$. The parameters $\bar{u}, \bar{v}$ will be called *admissible parameters* if the following conditions are satisfied.

1. The equations $u = u(\bar{u}, \bar{v})$, $v = v(\bar{u}, \bar{v})$ carry $\bar{u}^2 + \bar{v}^2 \leqq 1$ in a one-to-one and continuous way into $u^2 + v^2 \leqq 1$.

2. The functions $u(\bar{u}, \bar{v})$, $v(\bar{u}, \bar{v})$ have continuous partial derivatives of the first and second order for $\bar{u}^2 + \bar{v}^2 < 1$.

3. The JACOBIAN $\partial(u, v)/\partial(\bar{u}, \bar{v})$ is different from zero for $\bar{u}^2 + \bar{v}^2 < 1$.

Changing to admissible parameters $\bar{u}, \bar{v}$, we obtain the equations of S in the form

$$S : x = \bar{x}(\bar{u}, \bar{v}), \quad y = \bar{y}(\bar{u}, \bar{v}), \quad z = \bar{z}(\bar{u}, \bar{v}), \quad \bar{u}^2 + \bar{v}^2 \leqq 1.$$

The representations of S, obtained in this way, will be called *admissible representations*.

For every admissible representation $S : \mathfrak{x} = \mathfrak{x}(u, v)$, $u^2 + v^2 \leqq 1$, consider the integral

$$J(R) = \tfrac{1}{2} \iint\limits_{u^2 + v^2 < 1} (E + G)\, du\, dv,$$

where R denotes the admissible representation which has been used to compute the integral.

This functional $J(R)$ is defined by the same integral as the functional $A(T)$ of J. DOUGLAS (see V.13). It should be remembered, however, that the argument of $A(T)$ is a variable representation T of a given JORDAN curve, while the argument R of $J(R)$ is a variable representation of a given surface.

Suppose now that for the given surface S the greatest lower bound of $J(R)$ is finite. Then we have the following

Lemma. *If the variation problem $J(R) = minimum$ has a solution R, then R is an isothermic representation of S. Conversely, if S admits of an isothermic representation R, then R is a solution of the variation problem $J(R) = minimum$.*

The second half of the lemma is trivial. Indeed, we have for every admissible representation R the relations (5.12). Hence $\mathfrak{A}(S) \leqq J(R)$ for every admissible representation, while for isothermic representations we have obviously $\mathfrak{A}(S) = J(R)$.

The first half of the lemma would also be trivial if we would know a priori that S admits of isothermic representations. Indeed, if R_0 is an isothermic representation, then $\mathfrak{A}(S) = J(R_0)$, while $\mathfrak{A}(S) \leqq J(R)$

for all admissible representations. Hence $\min J(R) = \mathfrak{A}(S)$. If then a representation R minimizes $J(R)$, we must have $\mathfrak{A}(S) = J(R)$. It follows that in the relations (5.12) we must have the sign of equality all over, which obviously implies that $E = G$, $F = 0$.

The point therefore is that the lemma is true without any previous assumption concerning the existence of isothermic representations. The proof of the lemma runs as follows. Let

$$S : x = x(u, v), \quad y = y(u, v), \quad z = z(u, v), \quad u^2 + v^2 \leqq 1$$

be the minimizing representation. Let $\alpha = \alpha(u, v)$, $\beta = \beta(u, v)$ be admissible parameters, and let

$$S : x = \bar{x}(\alpha, \beta), \quad y = \bar{y}(\alpha, \beta), \quad z = \bar{z}(\alpha, \beta), \quad \alpha^2 + \beta^2 \leqq 1$$

be the resulting admissible representation $\bar{R}$. Then $J(\bar{R})$ is an integral over $\alpha^2 + \beta^2 < 1$. Introducing in this integral the old parameters u, v as variables of integration, we obtain the formula

$$J(\bar{R}) = \frac{1}{2} \iint\limits_{u^2 + v^2 < 1} [E(\alpha_v^2 + \beta_v^2) - 2F(\alpha_u \alpha_v + \beta_u \beta_v) + G(\alpha_u^2 + \beta_u^2)] \frac{\partial(u, v)}{\partial(\alpha, \beta)} \, du \, dv. \quad (5.13)$$

Let ε be a small parameter, and let $\psi(u, v)$ be a function which is continuous and has continuous partial derivatives of the first and second order in and on the unit circle $u^2 + v^2 = 1$. Then the equations

$$\alpha = u \cos \varepsilon \psi - v \sin \varepsilon \psi,$$
$$\beta = u \sin \varepsilon \psi + v \cos \varepsilon \psi$$

define admissible parameters for small values of ε, as it is easily seen. With these parameters in (5.13), $J(\bar{R})$ becomes a function $f(\varepsilon)$ of ε, which has a minimum for $\varepsilon = 0$. The equation $f'(0) = 0$ gives the relation

$$\iint \{[v(E - G) - 2uF] \psi_u + [u(E - G) + 2vF] \psi_v\} du \, dv = 0 . \quad (5.14)$$

Let now $\varphi(u, v)$ be a function having the properties required of $\psi(u, v)$ and having the additional property of vanishing on $u^2 + v^2 = 1$. The formulas[1]

$$\alpha = u + \varepsilon \varphi(u, v), \quad \beta = v$$

define then admissible parameters for small values of ε. Using these parameters in (5.13), we obtain

$$\iint [(E - G) \varphi_u + 2F \varphi_v] du \, dv = 0 . \quad (5.15)$$

The equations (5.14) and (5.15) are both of the form

$$\iint [a(u, v) \lambda_u + b(u, v) \lambda_v] du \, dv = 0 ,$$

where $a(u, v)$, $b(u, v)$ have continuous first partial derivatives in $u^2 + v^2 < 1$. The function $\lambda(u, v)$ has to be sufficiently regular and

---

[1] We are using here again the method of the variation of the independent variables, which also played an important part in the non-parametric problem (see IV.12).

has to vanish on $u^2 + v^2 = 1$.[1] This is exactly the situation which arises in the discussion of the first variation of multiple integrals; the classical inference is that $a_u + b_v = 0$. Applying this to (5.14) and (5.15), we obtain[2]

$$\frac{\partial}{\partial u}\left[v(E - G) - 2uF\right] + \frac{\partial}{\partial v}\left[u(E - G) + 2vF\right] = 0, \quad (5.16)$$

$$\frac{\partial}{\partial u}(E - G) + \frac{\partial}{\partial v}(2F) = 0, \quad (5.17)$$

for $u^2 + v^2 < 1$. From (5.16) follows the existence, in $u^2 + v^2 < 1$, of a function $\Omega(u, v)$ such that

$$\Omega_v = v(E - G) - 2uF, \quad \Omega_u = -u(E - G) - 2vF. \quad (5.18)$$

From (5.16), (5.17), (5.18) it follows by simple computations that $\Omega_{uu} + \Omega_{vv} = 0$, that is to say that Ω is harmonic. (5.14) can then be written in the form

$$\int\int(\Omega_v \psi_u - \Omega_u \psi_v)\,du\,dv = 0,$$

where ψ is *not* supposed to vanish on $u^2 + v^2 = 1$. If we substitute for the harmonic function Ω its FOURIER expansion and if we put $\psi = r^n \cos n\Theta$ and $\psi = r^n \sin n\Theta$, where $r\cos\Theta = u$, $r\sin\Theta = v$, it follows immediately that Ω is constant. Hence $\Omega_u = 0$, $\Omega_v = 0$, and (5.18) gives the desired equations $E = G$, $F = 0$.

V.26. To apply the preceding result to the functional $A(T)$ of J. DOUGLAS, it is necessary to restate the problem $A(T) = $ minimum in the following form. Let Γ^* be a given JORDAN curve. Consider all surfaces S which admit of a representation

$$S: x = x(u, v), \quad y = y(u, v), \quad z = z(u, v), \quad u^2 + v^2 \leq 1,$$

with the following properties.

1. $x(u, v)$, $y(u, v)$, $z(u, v)$ have continuous partial derivatives of the first and second order in $u^2 + v^2 < 1$.

2. $x(u, v)$, $y(u, v)$, $z(u, v)$ are continuous in $u^2 + v^2 \leq 1$, and the equations $x = x(u, v)$, $y = y(u, v)$, $z = z(u, v)$ define a monotonic transformation of $u^2 + v^2 = 1$ into a set on Γ^* (this implies that this set contains at least three distinct points, and on account of the assumed continuity it follows that the set covers the whole curve Γ^*).

We shall say that S is an *admissible surface, given in an admissible representation*. Let then the functional $J(S; R)$ be defined by

$$J(S; R) = \tfrac{1}{2}\int\int(E + G),$$

where R is the admissible representation which has been used to compute the integral.

The variation problem $J(S; R) = $ minimum is then clearly equivalent to the variation problem $A(T) = $ minimum of DOUGLAS.

---

[1] (5.14) holds even if λ does not vanish on the boundary.

[2] The lemma of HAAR (see IV.5) would permit to extend the proof to surfaces of class C'.

The application of the lemma in V.25 is now obvious. If R_0 is an admissible representation of an admissible surface S_0 such that $J(S_0;R_0)$ = minimum, that is to say such that $J(S_0; R_0) \leq J(S; R)$ for all admissible representations of all the admissible surfaces, then in particular $J(S_0; R_0) \leq J(S_0; R)$ for all the admissible representations R of S_0 itself. Hence R_0 is an isothermic representation, on account of the lemma in V.25.

Chapter VI.

The simultaneous problem in the parametric form. Generalizations.

VI.1. This problem has been investigated in the following statement. Given, in the xyz-space, a JORDAN curve Γ^*, consider all the continuous surfaces, of the type of the circular disc (see I.8), bounded by Γ^*, and suppose that the greatest lower bound $\mathfrak{a}(\Gamma^*)$ of their areas is finite. *Determine a solution S of problem P_2 (see III.5), such that $\mathfrak{A}(S) = \mathfrak{a}(\Gamma^*)$.*

While the method of GARNIER (see V.1) does not yield any information whatsoever as to the minimizing property of the solution obtained, both T. RADÓ and J. DOUGLAS proved that their respective methods yield a solution the area of which is a minimum. In either case, the proof depends essentially upon the existence theorem for conformal maps of polyhedrons, which has been the essential tool in the method reviewed in V.20 to V.23. These proofs will now be described briefly.

VI.2. Let $m(\Gamma^*)$ denote the smallest possible limit of the areas of sequences of polyhedrons $\mathfrak{P}_n$, such that the boundary polygon of $\mathfrak{P}_n$ converges, in the FRÉCHET sense, to Γ^*. This quantity $m(\Gamma^*)$ has been called by LEBESGUE, who introduced it in his Thesis[1], *the minimum area of Γ^**. Supposing that $m(\Gamma^*)$ is finite, T. RADÓ[2] starts with the following *approximate problem*. Take three distinct points A, B, C on the unit circle $u^2 + v^2 = 1$, three distinct points A^*, B^*, C^* on Γ^*, and give an $\varepsilon > 0$. Determine three functions $x(u, v), y(u, v), z(u, v)$ with the following properties.

1. $x(u, v), y(u, v), z(u, v)$ are harmonic in $u^2 + v^2 < 1$, and
2. satisfy the inequalities

$$\iint\limits_{u^2+v^2<1} (E^{\frac{1}{2}} - G^{\frac{1}{2}})^2 \, du \, dv \leq \varepsilon, \quad \iint\limits_{u^2+v^2<1} |F| \, du \, dv \leq \varepsilon.$$

3. $x(u, v), y(u, v), z(u, v)$ are continuous in $u^2 + v^2 \leq 1$, and the equations $x = x(u, v), y = y(u, v), z = z(u, v)$ carry $u^2 + v^2 = 1$ in a one-to-one and continuous way into a (not prescribed) JORDAN curve

[1] Intégrale, longueur, aire. Ann. Mat. pura appl. Vol. 7 (1902) pp. 231−359.
[2] The problem of the least area and the problem of PLATEAU. Math. Z. Vol. 32 (1930) pp. 763−796.

Γ_ε^*, the distance of which from Γ^* is $\leq \varepsilon$. Furthermore, the points A, B, C are carried into three points $A_\varepsilon^*, B_\varepsilon^*, C_\varepsilon^*$ such that the distances $A^* A_\varepsilon^*$, $B^* B_\varepsilon^*$, $C^* C_\varepsilon^*$ are all three $\leq \varepsilon$.

$$4. \iint\limits_{u^2+v^2<1} (EG - F^2)^{\frac{1}{2}}\, du\, dv \leq m(\Gamma^*) + \varepsilon .$$

The solution of this approximate problem is obtained by starting with a polyhedron $\mathfrak{P}$ with a boundary polygon $\mathfrak{p}$ such that both the difference $\mathfrak{A}(\mathfrak{P}) - m(\Gamma^*)$ and the distance $d(\mathfrak{p}, \Gamma^*)$ are less than a properly chosen $\sigma > 0$. If $x = \tilde{x}(u, v)$, $y = \tilde{y}(u, v)$, $z = \tilde{z}(u, v)$, $u^2 + v^2 \leq 1$, is a properly normalized isothermic representation of $\mathfrak{P}$, then the harmonic functions $x(u, v)$, $y(u, v)$, $z(u, v)$, which coincide with $\tilde{x}(u, v)$, $\tilde{y}(u, v)$, $\tilde{z}(u, v)$ on $u^2 + v^2 = 1$, solve the approximate problem. The proof follows by a slight and obvious modification of the reasoning described in V.22.

VI.3. Keep then the points A, B, C, A^*, B^*, C^* fixed and solve the preceding approximate problem for $\varepsilon = \varepsilon_n$, $\varepsilon_n \to 0$. Denote by $x_n(u, v)$, $y_n(u, v)$, $z_n(u, v)$ the solution. A slight modification of the passage to the limit, used in V.23, shows that there exists a subsequence which converges to a solution of the exact problem, as stated in VI.1. Again, the approximate conditions

$$\int\int (E_n^{\frac{1}{2}} - G_n^{\frac{1}{2}})^2 \leq \varepsilon_n , \qquad \iint |F| \leq \varepsilon_n ,$$

prove to be just as effective as the exact conditions $E = G$, $F = 0$ would be. A few remarks are necessary in connection with the effect of condition 4 of the approximate problem. The immediate information, obtained from the passage to the limit, concerning the area $\mathfrak{A}(S)$ of the limit surface

$$S : x = x(u, v), \quad y = y(u, v), \quad z = z(u, v), \quad u^2 + v^2 \leq 1$$

is that $\mathfrak{A}(S) = m(\Gamma^*)$. However, since S is bounded by Γ^*, we have $\mathfrak{A}(S) \leq \mathfrak{a}(\Gamma^*)$ by the definition of $\mathfrak{a}(\Gamma^*)$, while, on the other hand, obviously $m(\Gamma^*) \leq \mathfrak{a}(\Gamma^*)$. Consequently $\mathfrak{a}(\Gamma^*) = m(\Gamma^*) = \mathfrak{A}(S)$, that is to say the area of S is a minimum. It also follows that $\mathfrak{a}(\Gamma^*) = m(\Gamma^*)$, whenever $m(\Gamma^*)$ is finite. If $m(\Gamma^*) = +\infty$, then on account of $m(\Gamma^*) \leq \mathfrak{a}(\Gamma^*)$ we also have $\mathfrak{a}(\Gamma^*) = +\infty$. Hence, for every JORDAN curve Γ^*, we have $\mathfrak{a}(\Gamma^*) = m(\Gamma^*)$[1]. Thus the condition $m(\Gamma^*) < +\infty$, under which the approximate problem has been solved in VI.2, is equivalent to the condition $\mathfrak{a}(\Gamma^*) < +\infty$. It follows that *the simultaneous problem is solvable for every* JORDAN *curve which bounds some continuous surface, of the type of the circular disc, with a finite area.* This condition obviously is satisfied if Γ^* is rectifiable. It is also clearly

---

[1] This generalizes a result of LEBESGUE: Intégrale, longueur, aire. Ann. Mat. pura appl. Vol. 7 (1902) pp. 231—359.

satisfied if Γ^* is situated on a closed convex surface (since the area of such a surface is always finite and consequently the areas of the two regions, bounded on the surface by Γ^*, are also finite).

VI.4. Both T. RADÓ and J. DOUGLAS observe that for a general JORDAN curve Γ^* the condition $\mathfrak{a}(\Gamma^*) < +\infty$ is not satisfied. While this fact is rather obvious, the example given by J. DOUGLAS[1] is not quite convincing. J. DOUGLAS constructs a JORDAN curve Γ^* such that the measure of the orthogonal projection of Γ^* upon the xy-plane is $+\infty$, provided every point of the projection is counted with the proper multiplicity. From this it should follow that for every continuous surface S, of the type of the circular disc, the area $\mathfrak{A}(S)$ must also be $+\infty$, a conclusion which certainly is not obvious as it stands, since $\mathfrak{A}(S)$, in general, is *smaller* than the measure of the orthogonal projection of S upon a plane[2]. The reviewer takes the liberty to point out a simple way of obtaining more satisfactory examples. The orthogonal projection, upon the xy-plane, of a JORDAN curve Γ^* is a closed continuous curve C^*. Every point (x, y), not on C^*, has a definite integer $n(x, y)$ as its topological index with respect to C^*. Define $N(x, y)$ by $N(x, y) = |n(x, y)|$ if (x, y) is not on C^* and by $N(x, y) = 0$ if (x, y) is on C^*. Then $N(x, y)$ is a measurable function which is clearly zero outside of a sufficiently large circle K. Furthermore, $\mathfrak{A}(S) \geqq \iint N(x, y) dx dy$† for every continuous surface, of the type of the circular disc, bounded by Γ^*, and consequently $\mathfrak{a}(\Gamma^*) \geqq \iint N(x, y) dx dy$. It is clear on the other hand that if the projection C^* of Γ^* contains a properly arranged spiral-shaped arc, $\iint N(x, y) dx dy$ can be driven up to $+\infty$.

VI.5. To prove the minimizing property of the minimal surface obtained by his method, J. DOUGLAS[3] first observes that if m denotes the (by assumption finite) greatest lower bound of the functional $A(T)$ for the given JORDAN curve Γ^*, then $\mathfrak{A}(S) \geqq m$ for every continuous surface, of the type of the circle, bounded by Γ^*. Let us observe that this would be obvious if S would admit of an isothermic representation $\mathfrak{x} = \mathfrak{x}(u, v)$, $u^2 + v^2 \leqq 1$, $E = G$, $F = 0$. Indeed, denote by T the transformation of $u^2 + v^2 = 1$ into Γ^* defined by the equation of the surface and denote by $\tilde{x}(u, v)$, $\tilde{y}(u, v)$, $\tilde{z}(u, v)$ the harmonic functions coinciding with the components $x(u, v)$, $y(u, v)$, $z(u, v)$ of $\mathfrak{x}$ on $u^2 + v^2 = 1$. Then

$$A(T) = \tfrac{1}{2} \int\!\int (\tilde{E} + \tilde{G}).$$

---

[1] Solution of the problem of PLATEAU. Trans. Amer. Math. Soc. Vol. 33 (1931) pp. 320—321.

[2] See I.5 and I.6.

† This follows easily from the theorems in T. RADÓ: Über das Flächenmaß rektifizierbarer Flächen. Math. Ann. Vol. 100 (1928) pp. 445—479 § 2.

[3] Solution of the problem of PLATEAU. Trans. Amer. Math. Soc. Vol. 33 (1931) pp. 318—321. See also the abstract, Existence of a surface etc. Bull. Amer. Math. Soc. Vol. 36 (1930) pp. 796—797.

On account of the minimizing property of harmonic functions, we also have (cf. II.10)
$$\tfrac{1}{2}\int\int(\tilde{E}+\tilde{G})\leqq\tfrac{1}{2}\int\int(E+G)\,.$$
On account of $E=G$, $F=0$, we have
$$\mathfrak{A}(S)=\tfrac{1}{2}\int\int(E+G)\,.$$
Finally $m\leqq A(T)$ by the definition of m. Consequently $m\leqq\mathfrak{A}(S)$.

It is then easy to understand the proof of J. Douglas for a general S. The surface is approximated by polyhedrons and the inequality $m\leqq\mathfrak{A}(S)$ follows then, by several passages to the limit, from the fact that these approximating polyhedrons do admit of isothermic representations. While it follows from this reasoning that $m\leqq a(\varGamma^*)$, it is on the other hand obvious that we have, for the minimal surface S obtained by the method of Douglas, $m=\mathfrak{A}(S)$, which implies $m\geqq a(\varGamma^*)$. Hence $\mathfrak{A}(S)=m=a(\varGamma^*)$. Thus it follows that S has the minimizing property.

It follows in this manner also that *the assumption $m<+\infty$, under which the method of J. Douglas operates, is equivalent to the assumption that $\varGamma^*$ bounds some continuous surface, of the type of the circular disc, with a finite area.*

It is strange that while the method of Douglas is otherwise independent of the theory of conformal mapping, he needs this theory to determine the generality of the method and also to establish the fact that the solution has a minimum area. It should be observed however that J. Douglas considers this part of his work as a stop-gap, and expects to develop a method entirely independent of the theory of conformal mapping[1].

VI.6. J. Douglas gave another solution of the simultaneous problem (see VI.1) on the following basis[2]. Take a sequence of polyhedrons $\mathfrak{P}_n$, with boundary polygons converging to the given Jordan curve $\varGamma^*$, such that $\mathfrak{A}(\mathfrak{P}_n)$ converges to the (by assumption finite) minimum area $m(\varGamma^*)$ of $\varGamma^*$. Let $\mathfrak{P}_n: \mathfrak{x}=\mathfrak{x}_n(u,v)$, $u^2+v^2\leqq1$, be an isothermic representation of $\mathfrak{P}_n$. These equations define a transformation T_n of $u^2+v^2=1$ into the boundary polygon $\mathfrak{p}_n$ of $\mathfrak{P}_n$, and if the isothermic representations $\mathfrak{x}=\mathfrak{x}_n(u,v)$ are properly normalized, then the sequence T_n will contain a subsequence converging to a transformation
$$T: x=\xi(\Theta)\,,\quad y=\eta(\Theta)\,,\quad z=\zeta(\Theta)$$
of the unit circle $u=\cos\Theta$, $v=\sin\Theta$ into $\varGamma^*$, such that the harmonic functions $x(u,v),y(u,v),z(u,v)$ with the boundary values $\xi(\Theta),\eta(\Theta),\zeta(\Theta)$ solve the simultaneous problem for $\varGamma^*$. Since the triples of harmonic

---

[1] See J. Douglas: Solution of the problem of Plateau. Trans. Amer. Math. Soc. Vol. 33 (1931) p. 265.

[2] J. Douglas: The mapping theorem of Koebe and the problem of Plateau. J. Math. Physics, Massachusetts Inst. Technol. Vol. 10 (1931) pp. 106—130.

functions $\tilde{x}_n(u, v)$, $\tilde{y}_n(u, v)$, $\tilde{z}_n(u, v)$, which coincide on $u^2 + v^2 = 1$ with the components $x_n(u, v)$, $y_n(u, v)$, $z_n(u, v)$ of $\mathfrak{x}_n(u, v)$, are obviously identical to the approximate solutions used in VI.2, this method differs, both in its general layout and in its details, very little from the one reviewed in VI.2 to VI.3.

VI.7. For JORDAN curves Γ^* such that $m(\Gamma^*) = +\infty$, J. DOUGLAS obtained, by a passage to the limit, the theorem that *problem P_2* (see III.5) *admits of a solution S such that every interior portion of S has a minimum area with respect to its own boundary curve*[1]. The proof proceeds as follows. Take a sequence $\mathfrak{p}_n^*$ of simple closed polygons converging, in the sence of FRÉCHET, to Γ^*. For $\mathfrak{p}_n^*$, the simultaneous problem is certainly solvable; let

$$S_n : x = x_n(u, v), \quad y = y_n(u, v), \quad z = z_n(u, v), \quad u^2 + v^2 \leq 1, \quad (6.1)$$

be a solution, normalized by the condition that three distinct fixed points A, B, C on $u^2 + v^2 = 1$ are carried into three distinct points A_n^*, B_n^*, C_n^* of $\mathfrak{p}_n^*$ which converge to three distinct points A^*, B^*, C^* on Γ^*. On account of the approximation theorem (see V.9), it is legitimate to suppose that the sequence converges, in $u^2 + v^2 \leq 1$, to a solution

$$S : x = x(u, v), \quad y = y(u, v), \quad z = z(u, v), \quad u^2 + v^2 \leq 1$$

of problem P_2 for Γ^*. Since all the functions involved are harmonic and uniformly bounded, the convergence extends to partial derivatives and is uniform in every smaller concentric circle. Denote by $S^{(r)}$ the portion of S corresponding to $u^2 + v^2 \leq r^2 < 1$. J. DOUGLAS proves then that $S^{(r)}$ has a minimum area, with respect to its own boundary curve, by computations whose geometrical background might be explained as follows. Denote by $l^{(r)}$ the length of the boundary curve of $S^{(r)}$, and let $S_n^{(r)}$, $l_n^{(r)}$ have similar meanings for S_n. Since all the partial derivatives converge uniformly in $u^2 + v^2 \leq r^2 < 1$, it follows that $l_n^{(r)} \to l^{(r)}$, $\mathfrak{A}(S_n^{(r)}) \to \mathfrak{A}(S^{(r)})$. It is then easily seen that we have a ring-shaped surface $\Sigma_n^{(r)}$, bounded by the boundary curves of $S_n^{(r)}$ and $S^{(r)}$, such that $\mathfrak{A}(\Sigma_n^{(r)}) \to 0$.

Suppose there exists a continuous surface $\overline{S}^{(r)}$, of the type of the circular disc, with the same boundary as $S^{(r)}$ and such that $\mathfrak{A}(\overline{S}^{(r)}) \leq \mathfrak{A}(S^{(r)}) - \varepsilon$, $\varepsilon > 0$. Denote by $\overline{S}_n$ the surface $\overline{S}^{(r)} + \Sigma_n^{(r)}$. Then $\mathfrak{A}(\overline{S}_n^{(r)}) \to \mathfrak{A}(\overline{S}^{(r)})$, and from the relations $\mathfrak{A}(\overline{S}^{(r)}) \leq \mathfrak{A}(S^{(r)}) - \varepsilon$, $\mathfrak{A}(S_n^{(r)}) \to \mathfrak{A}(S^{(r)})$ it follows that $\mathfrak{A}(\overline{S}_n^{(r)}) < \mathfrak{A}(S_n^{(r)})$ for large values of n. Replacing then, for S_n, the portion $S_n^{(r)}$ by the surface $\overline{S}_n^{(r)}$, we obtain a surface with area less than S_n, in contradiction with the minimizing property of S_n.

---

[1] J. DOUGLAS: The least area property of the minimal surface determined by an arbitrary JORDAN contour. Proc. Nat. Acad. Sci. U. S. A. Vol. 17 (1931) pp. 211—216.

Thus $S^{(r)}$ has a minimum area with respect to its own boundary, and consequently every portion of $S^{(r)}$ has the same property. Since r was arbitrary, this proves the theorem.

VI.8. The reviewer has had the privilege of seeing the manuscript of a paper by J. E. McShane[1], in which the solution of the simultaneous problem is derived from general theorems, interesting in themselves.

The following definitions are used by McShane. A function $f(u, v)$, given in $u^2 + v^2 \leq 1$, will be said to satisfy the *condition* (C) if

1. $f(u, v)$ is continuous in $u^2 + v^2 \leq 1$, and

2. $f(u, v)$ is for almost every u an absolutely continuous function of v, and for almost every v an absolutely continuous function of u, and

3. the Dirichlet integral $\int\int (f_u^2 + f_v^2)$, extended over $u^2 + v^2 < 1$, is finite.

Let again $f(u, v)$ be continuous in $u^2 + v^2 \leq 1$. Let R be a sub-region of $u^2 + v^2 \leq 1$, and denote by $M(R)$, $m(R)$ the maximum and minimum of $f(u, v)$ in R, and by $M_b(R)$, $m_b(R)$ the maximum and minimum of $f(u, v)$ on the boundary of R. The least upper bound of $M(R) - M_b(R)$, $m_b(R) - m(R)$, for all subregions R of $u^2 + v^2 \leq 1$, will be called the *monotonic deficiency* of $f(u, v)$ in $u^2 + v^2 \leq 1$. Clearly, $f(u, v)$ is monotonic, is the sense explained in IV.18, if and only if the monotonic deficiency is zero.

VI.9. With these definitions we have the following *selection theorem*. Let $f_n(u, v)$ be a sequence of functions in $u^2 + v^2 \leq 1$, such that 1. the Dirichlet integrals, extended over $u^2 + v^2 < 1$ are uniformly bounded, 2. every $f_n(u, v)$ satisfies condition (C), 3. the sequence converges uniformly on $u^2 + v^2 = 1$, and 4. the monotonic deficiency of $f_n(u, v)$ converges to zero for $n \to \infty$. *Then the sequence contains a subsequence which converges uniformly in $u^2 + v^2 \leq 1$, and the limit function satisfies again condition (C).*

Except for the second half of the theorem, concerned with the condition (C), this theorem is a generalization of a similar selection theorem used by Lebesgue[2]. As observed by McShane, the proof of Lebesgue extends easily to this more general case. Condition (C) also can easily be handled on the basis of general theorems in the theory of functions of real variables.

VI.10. Another *selection theorem* used by McShane is concerned with *surfaces*. Let

$$S_n : x = x_n(u, v), \quad y = y_n(u, v), \quad z = z_n(u, v), \quad u^2 + v^2 \leq 1 \quad (6.2)$$

be a sequence of surfaces with the following properties.

---

[1] See the abstract: J. E. McShane: Parametrization of saddle-surfaces, with application to the problem of Plateau. Bull. Amer. Math. Soc. Vol. 38 (1932) pp. 810—811. The detailed presentation appears in the Trans. Amer. Math. Soc.

[2] Sur le problème de Dirichlet. Rend Circ. mat. Palermo Vol. 24 (1907) pp. 371—402.

1. All the coordinate functions satisfy condition (C) of VI.8.

2. The DIRICHLET integrals, over $u^2 + v^2 < 1$, of the coordinate functions are less than a fixed constant M independent of n.

3. The monotonic deficiency of $x_n(u, v), y_n(u, v), z_n(u, v)$ converges to zero.

4. The equations (6.2) take $u^2 + v^2 = 1$ in a one-to-one and continuous way into a JORDAN curve Γ_n^*. These curves Γ_n^* converge, in the sense of FRÉCHET, to a JORDAN curve Γ^*. There exist three distinct points A, B, C on $u^2 + v^2 = 1$ whose images A_n^*, B_n^*, C_n^* converge to three distinct points A^*, B^*, C^* of Γ^*.

Then the sequence (6.2) contains a subsequence which converges uniformly in $u^2 + v^2 \leq 1$. The limit surface

$$S : x = x(u, v), \quad y = y(u, v), \quad z = z(u, v); \quad u^2 + v^2 \leq 1, \quad (6.2.1)$$

is such that $x(u, v), y(u, v), z(u, v)$ are monotonic and satisfy condition (C). Furthermore, the equations (6.2.1) define, for $u^2 + v^2 = 1$, a continuous monotonic transformation of $u^2 + v^2 = 1$ into Γ^*, such that A, B, C are carried into A^*, B^*, C^*.

To prove this, denote by T_n the monotonic transformation of $u^2 + v^2 = 1$ into Γ_n^*, defined by (6.2) for $u^2 + v^2 = 1$. All the conditions for the selection theorem in I.22 being satisfied, there is a convergent subsequence, to be denoted again by T_n, which converges to a monotonic transformation T of $u^2 + v^2 = 1$ into Γ^*. The continuity of T follows then from assumption 2 by a reasoning similar to the lemma in V.15.

Now then, if a sequence of monotonic functions converges, on a closed interval, to a continuous function, then the convergence necessarily is uniform[1]. McSHANE observes that this obviously remains true for sequences of monotonic transformations. From the continuity of T it follows therefore that T_n converges uniformly to T; in other words, the sequence (6.2) contains a subsequence which converges uniformly on $u^2 + v^2 = 1$. The present selection theorem is then an immediate consequence of the selection theorem in VI.9.

VI.11. A continuous surface

$$S : x = x(u, v), \quad y = y(u, v), \quad z = z(u, v), \quad u^2 + v^2 \leq 1 \quad (6.3)$$

is called by McSHANE a *saddle-surface* if $x(u, v), y(u, v), z(u, v)$ are monotonic. He proves that this property is independent of the particular choice of the parametric representation of the surface. His main theorem is then that *every saddle-surface which has a finite area and which is bounded by a* JORDAN *curve admits of isothermic parameters in the following generalized sense.*

---

[1] See for instance PÓLYA-SZEGÖ: Aufgaben und Lehrsätze Vol. 1 p. 63 problem 127.

If S is a saddle-surface, bounded by a JORDAN curve Γ^*, with finite area, then S admits of a representation (6.3) with the following properties.

1. $x(u, v)$, $y(u, v)$, $z(u, v)$ satisfy condition (C) of VI.8 and are monotonic functions.

2. $E = G$, $F = 0$ almost everywhere in $u^2 + v^2 < 1$.

The proof proceeds in the following steps. By the definition of $\mathfrak{A}(S)$, we have a sequence of polyhedrons $\mathfrak{P}_n$ such that $\mathfrak{P}_n \to S$ and $\mathfrak{A}(\mathfrak{P}_n) \to \mathfrak{A}(S)$. Choose three distinct fixed points A, B, C on $u^2 + v^2 = 1$, three distinct fixed points A^*, B^*, C^* on the boundary curve Γ^* of S, and three distinct points A_n^*, B_n^*, C_n^* on the boundary polygon of $\mathfrak{P}_n$, such that $A_n^* \to A^*$, $B_n^* \to B^*$, $C_n^* \to C^*$ (this is possible on account of $\mathfrak{P}_n \to S$). Let

$$\mathfrak{P}_n : x = x_n(u, v), \quad y = y_n(u, v), \quad z = z_n(u, v), \quad u^2 + v^2 \leq 1 \quad (6.4)$$

be an isothermic representation of $\mathfrak{P}_n$ (see I.19), such that A, B, C are carried into A_n^*, B_n^*, C_n^*. Then, on account of $E_n = G_n$, $F_n = 0$ we have

$$\mathfrak{A}(\mathfrak{P}_n) = \tfrac{1}{2}\iint (E_n + G_n). \quad (6.5)$$

Since $\mathfrak{A}(\mathfrak{P}_n) \to \mathfrak{A}(S)$, and $\mathfrak{A}(S)$ is finite, the integrals (6.5) are uniformly bounded, and consequently the DIRICHLET integrals of $x_n(u, v)$, $y_n(u, v)$, $z_n(u, v)$ are less than a finite constant M independent of n. From $\mathfrak{P}_n \to S$ it follows, on account of the assumption that S is a saddle-surface, that the monotonic deficiencies of $x_n(u, v)$, $y_n(u, v)$, $z_n(u, v)$ converge to zero. Finally, the isothermic representation (6.4) of $\mathfrak{P}_n$ clearly satisfies condition (C) of VI.8. Thus all the assumptions of the selection theorem of VI.10 are satisfied, and therefore the sequence (6.4) contains a subsequence converging uniformly in $u^2 + v^2 \leq 1$. For the sake of simplicity, let (6.4) denote such a subsequence. If $x^*(u, v)$, $y^*(u, v)$, $z^*(u, v)$ denote the limit functions, then

$$S : x = x^*(u, v), \quad y = y^*(u, v), \quad z = z^*(u, v), \quad u^2 + v^2 \leq 1$$

is a representation[1] of S as required by the theorem. Indeed, 1. is satisfied on account of the selection theorem of VI.10. To verify 2., observe that on account of I.16, we have

$$\mathfrak{A}(S) = \iint (E^* G^* - F^{*2})^{\frac{1}{2}}.$$

On the other hand, from $x_n(u, v) \rightrightarrows x_n^*(u, v)$ it follows, on account of the lower semi-continuity of the DIRICHLET integral[2], that

$$\iint (x_u^{*2} + x_v^{*2}) \leq \underline{\lim} \iint (x_{nu}^2 + x_{nv}^2).$$

Write the corresponding inequalities for the y and z coordinates and add. There follows

$$\tfrac{1}{2}\iint (E^* + G^*) \leq \lim \tfrac{1}{2}\iint (E_n + G_n) = \lim \mathfrak{A}(\mathfrak{P}_n) = \mathfrak{A}(S),$$

since

$$\tfrac{1}{2}\iint (E_n + G_n) = \mathfrak{A}(\mathfrak{P}_n) \to \mathfrak{A}(S).$$

---

[1] In the sense of I.8.　　　　[2] See IV.20.

Thus we obtain the inequality

$$\tfrac{1}{2}\iint (E^* + G^*) \leq \mathfrak{A}(S) = \iint (E^*G^* - F^{*2})^{\tfrac{1}{2}}, \tag{6.6}$$

while on the other hand

$$\iint (E^*G^* - F^{*2})^{\tfrac{1}{2}} \leq \iint E^{*\tfrac{1}{2}} G^{*\tfrac{1}{2}} \leq \tfrac{1}{2}\iint (E^* + G^*). \tag{6.7}$$

From (6.6) and (6.7) it follows immediately that $E^* = G^*$, $F^* = 0$ almost everywhere.

VI.12. Let

$$S : x = x(u, v), \quad y = y(u, v), \quad z = z(u, v), \quad u^2 + v^2 \leq 1$$

be a continuous surface bounded by a JORDAN curve Γ^*, such that $x(u, v)$, $y(u, v)$, $z(u, v)$ satisfy condition (C). Then, on account of I.16, $\mathfrak{A}(S)$ is finite and is given by

$$\mathfrak{A}(S) = \iint (EG - F^2)^{\tfrac{1}{2}}.$$

McSHANE proves then that there exists a saddle-surface

$$S^* : x = x^*(u, v), \quad y = y^*(u, v), \quad z = z^*(u, v), \quad u^2 + v^2 \leq 1,$$

with the following properties.

1. S^* is again bounded by Γ^*,
2. $\mathfrak{A}(S^*) \leq \mathfrak{A}(S)$,
3. $x^*(u, v)$, $y^*(u, v)$, $z^*(u, v)$ satisfy condition (C) of VI.8. The construction of S^* is analogous to a construction used by LEBESGUE in the theory of the DIRICHLET problem[1]. Roughly speaking, if S is not itself a saddle-surface, then we shall have on S some closed plane curve $\overline{\Gamma}$ which bounds a portion $\overline{S}$ of S which is not in the plane of $\overline{\Gamma}$. If then we replace $\overline{S}$ by its orthogonal projection upon the plane of $\overline{\Gamma}$, we do not increase the area of S (condition (C) is necessary to justify this conclusion), and we bring S closer to being a saddle-surface. This operation, rigorously worded and applied a denumerable infinity of times, leads to the saddle-surface S^*.

VI.13. The preceding theorems lead to the following *selection theorem*. Let there be given a sequence

$$S_n : x = x_n(u, v), \quad y = y_n(u, v), \quad z = z_n(u, v), \quad u^2 + v^2 \leq 1,$$

of saddle-surfaces, such that every S_n is bounded by a JORDAN curve Γ_n^*. Suppose that the areas $\mathfrak{A}(S_n)$ are uniformly bounded, and that Γ_n^* converges, in the FRÉCHET sense, to a JORDAN curve Γ^*.

Then the sequence S_n contains a subsequence which converges in the sense of FRÉCHET. The limit surface is a saddle-surface bounded by Γ^*.

---

[1] LEBESGUE: Sur le problème de DIRICHLET. Rend. Circ. mat. Palermo Vol. 24 (1907) pp. 371—402.

The proof follows immediately from VI.12 and from the selection theorem of VI.10.

VI.14. On the basis of the preceding theorems, MᴄSʜᴀɴᴇ gives the following solution of the simultaneous problem (see VI.1). Denote by $\mathfrak{a}(\Gamma^*)$ the (by assumption finite) greatest lower bound of the areas of all continuous surfaces, of the type of the circular disc, bounded by Γ^*. Then first we havè a sequence of polyhedrons $\mathfrak{P}_n$ such that $\mathfrak{A}(\mathfrak{P}_n) \to \mathfrak{a}(\Gamma^*)$ and such that the boundary polygon $\mathfrak{p}_n^*$ of $\mathfrak{P}_n$ converges, in the Fʀéᴄʜᴇᴛ sense, to Γ^*. The theorem of VI.12 clearly applies to polyhedrons, hence we can replace $\mathfrak{P}_n$ by a saddle-surface S_n, bounded by $\mathfrak{p}_n^*$, such that $\mathfrak{A}(S_n) \leq \mathfrak{A}(\mathfrak{P}_n)$. Then clearly

$$\overline{\lim} \mathfrak{A}(S_n) \leq \lim \mathfrak{A}(\mathfrak{P}_n) = \mathfrak{a}(\Gamma^*). \qquad (6.8)$$

Thus the sequence $\mathfrak{A}(S_n)$ is bounded. Hence the selection theorem of VI.13 applies, and we obtain a surface S, limit of a subsequence of the sequence S_n, such that

1. S is bounded by Γ^*,
2. S is a saddle-surface,
3. $\mathfrak{A}(S) \leq \underline{\lim} \mathfrak{A}(S_n)$, on account of the lower semi-continuity of the area.

'From 3 and (6.8) it follows that $\mathfrak{A}(S) \leq \mathfrak{a}(\Gamma^*)$, while from 1 it follows that $\mathfrak{a}(\Gamma^*) \leq \mathfrak{A}(S)$. Hence $\mathfrak{A}(S) = \mathfrak{a}(\Gamma^*)$, that is to say S solves the least area problem for Γ^*. *While, in general, this fact alone would not permit any further conclusions* (see III.13), *the fact that S is a saddle-surface enables* MᴄSʜᴀɴᴇ *to show that S is a minimal surface.* Indeed, S being a saddle-surface with a finite area, it admits of an isothermic representation in the sense of the theorem of VI.11. Let

$$S : x = x(u, v), \quad y = y(u, v), \quad z = z(u, v), \quad u^2 + v^2 \leq 1 \quad (6.9)$$

be an isothermic representation of S, and denote by $\tilde{x}(u, v)$, $\tilde{y}(u, v)$, $\tilde{z}(u, v)$ the harmonic functions coinciding on $u^2 + v^2 = 1$ with $x(u, v)$, $y(u, v)$, $z(u, v)$. The reasoning of II.10 applies on account of condition (C) (see VI.8), and it follows that the area of the surface

$$\tilde{S} : x = \tilde{x}(u, v), \quad y = \tilde{y}(u, v), \quad z = \tilde{z}(u, v), \quad u^2 + v^2 \leq 1$$

is less than the area of S, unless $x(u, v)$, $y(u, v)$, $z(u, v)$ are themselves harmonic functions. Hence it follows from the minimizing property of S that the functions $x(u, v)$, $y(u, v)$, $z(u, v)$ in (6.9) are harmonic. Consequently, the equations $E = G$, $F = 0$, which are known to be satisfied almost everywhere in $u^2 + v^2 < 1$, will hold everywhere, since E, F, G are continuous (even analytic). That is to say, (6.9) solves the simultaneous problem for Γ^*.

VI.15. A little discussion is however necessary to finish up the existence proof. The selection theorem of VI.10, which plays an important part in the preceding proof, yielded a limit surface

$$S : x = x(u, v), \quad y = y(u, v), \quad z = z(u, v), \quad u^2 + v^2 \leq 1,$$

such that these equations define a continuous monotonic transformation T of $u^2 + v^2 = 1$ into the boundary curve Γ^*, three distinct points A, B, C given on $u^2 + v^2 = 1$ being carried into three distinct points A^*, B^*, C^* given on Γ^*. On the other hand, it does *not* follow from the proof of the selection theorem that T is a one-to-one transformation, and, as a consequence, the same remark applies to the solution of the simultaneous problem obtained by the method of McShane. This point can however be taken care of easily. If (6.9) represents the solution, then $x(u, v)$, $y(u, v)$, $z(u, v)$ are harmonic and $E = G$, $F = 0$ in $u^2 + v^2 < 1$. If then the equations do not carry $u^2 + v^2 = 1$ in a one-to-one way into Γ^*, then there exists (see I.23) an arc σ on $u^2 + v^2 = 1$ along which $x(u, v)$, $y(u, v)$, $z(u, v)$ all three reduce to constants. The reasoning at the end of V.10 shows then that $x(u, v)$, $y(u, v)$, $z(u, v)$ reduce to constants identically. This however contradicts the fact that there exists on $u^2 + v^2 = 1$ a triple of distinct points A, B, C which are carried by (6.9) into a triple of distinct points A^*, B^*, C^* of Γ^*.

It would be interesting to determine if, in the important theorem of VI.11 on the conformal maps of saddle-surfaces, the correspondence between the boundary curves is or is not necessarily one-to-one, although this mapping theorem is efficient enough as it stands, as far as the problem of Plateau is concerned.

VI.16. The existence of the solution of the simultaneous problem has several interesting implications, some of which will now be considered.

It has been observed (see III.14) that a minimal surface, bounded by a given curve, does not in general have a minimum area. H. A. Schwarz[1], by a discussion of the second variation, obtained conditions under which a given piece S of a minimal surface has a minimum area with respect to surfaces bounded by the same curve and sufficiently close to S (relative minimum). The recent results permit the establishing of the following theorem concerned with an absolute minimum[2].

If the boundary curve Γ^* of a minimal surface S has a simply-covered convex curve as its parallel or central projection upon some plane, then the area of S is a minimum with respect to all surfaces bounded by Γ^* (surface means continuous surface of the type of the circular disc).

First, on account of the remark at the end of VI.3, the minimum area of Γ^* is finite, and hence the simultaneous problem has a solution S^* (see VI.3). On account of the uniqueness theorem of III.11, S^* is identical to S, which proves the theorem.

VI.17. Let S denote a regular minimal surface (see III.4). A sufficiently small vicinity S_0 of any point P_0 of S has then a simply covered

[1] Gesammelte Mathematische Abhandlungen Vol. 1 pp. 151—167 and pp. 224 to 269.

[2] T. Radó: Contributions to the theory of minimal surfaces. Acta Litt. Sci. Szeged Vol. 6 (1932) pp. 1—20.

orthogonal projection upon, say, the xy-plane. Hence P_0 is contained in a portion S_0^* of S_0 such that the xy-projection of the boundary curve of S_0^* is a convex curve (a circle, if we wish). Hence, on account of VI.16, *the area of a minimal surface, in the sense of differential geometry, is an absolute minimum in the small*[1].

VI.18. Let us consider now the non-parametric problem (Chapter IV). Denote by C a convex JORDAN curve in the xy-plane, and by $\varphi(P)$ a continuous function of the point P varying on C. The equation $z = \varphi(P)$ determines, in the xyz-space, a JORDAN curve Γ^*, and C is the simply-covered orthogonal projection of Γ^*. Since C is convex, the simultaneous problem has a solution S for Γ^* (see the end of VI.3), and on account of III.10, S can be represented by an equation $z = z(x, y)$, where $z(x, y)$ is continuous in and on C, analytic in C, equal to $\varphi(P)$ on C, and satisfies in C the partial differential equation

$$(1 + q^2)r - 2pqs + (1 + p^2)t = 0.$$

That is to say, *the boundary value problem is solvable for every continuous boundary function $\varphi(P)$ given on a convex curve C.*[2] The three-point condition (see IV.2) is thus seen to be superfluous.

It also follows that the solution $z = z(x, y)$ defines a minimal surface whose area is a minimum with respect to all continuous surfaces, of the type of the circular disc, bounded by the same curve (and not only with respect to surfaces $z = z(x, y)$, where $z(x, y)$ satisfies the LIPSCHITZ condition, as it would follow from the existence proof described in IV.17 to IV.26). Thus $z(x, y)$ solves the problem $\mathfrak{A}(z) = \text{minimum}$ to be considered in the next section VI.19.

VI.19. Let C and $\varphi(P)$ have the same meaning as in the preceding section VI.18. Consider the totality of all those functions $z(x, y)$ which are continuous in and on C and are equal to $\varphi(P)$ on C. Denote by $\mathfrak{A}(z)$ the area, in the sense of LEBESGUE, of the surface $z = z(x, y)$. Consider the problem $\mathfrak{A}(z) = \text{minimum}$. On account of VI.18, the problem has a solution $z_1(x, y)$ (since C is supposed to be convex). McSHANE proved that the solution is *unique*[3]. To see this, observe first that

$$a = \min \mathfrak{A}(z)$$

is finite on account of the convexity of C (see the end of VI.3). Suppose we have another solution z_2. Then

$$\mathfrak{A}(z_1) = \mathfrak{A}(z_2) = a. \tag{6.10}$$

[1] T. RADÓ: Contributions to the theory of minimal surfaces. Acta Litt. Sci. Szeged Vol. 6 (1932) pp. 1—20.

[2] T. RADÓ: The problem of the least area and the problem of PLATEAU. Math. Z. Vol. 32 (1930) pp. 795—796. The theorem has been stated, without proof, by S. BERNSTEIN: Sur les équations du Calcul des Variations. Ann. École norm. Vol. 29 (1912) pp. 431—485. See in particular pp. 484—485.

[3] McSHANE: On a certain inequality of STEINER. Ann. of Math. Vol. 33 (1932) pp. 123—138.

Clearly, $\frac{1}{2}(z_1 + z_2)$ also reduces to $\varphi(P)$ on C. Hence

$$\mathfrak{A}\left(\frac{z_1 + z_2}{2}\right) \geq a,$$

while the inequality of STEINER (see I.17) gives

$$\mathfrak{A}\left(\frac{z_1 + z_2}{2}\right) \leq \tfrac{1}{2}\left(\mathfrak{A}(z_1) + \mathfrak{A}(z_2)\right) = a.$$

Thus

$$\mathfrak{A}\left(\frac{z_1 + z_2}{2}\right) = \tfrac{1}{2}\left(\mathfrak{A}(z_1) + \mathfrak{A}(z_2)\right). \tag{6.11}$$

Consequently, as shown by McSHANE (see I.17), $p_1 = p_2$, $q_1 = q_2$ almost everywhere (p, q denote the partial derivatives of the first order). Well then, z_1 is analytic and thus

$$\mathfrak{A}(z_1) = \iint (1 + p_1^2 + q_1^2)^{\frac{1}{2}}. \tag{6.12}$$

Since $p_1 = p_2$, $q_1 = q_2$ almost everywhere, we have

$$\iint (1 + p_1^2 + q_1^2)^{\frac{1}{2}} = \iint (1 + p_2^2 + q_2^2)^{\frac{1}{2}}. \tag{6.13}$$

From (6.10), (6.12) and (6.13) it follows that

$$\mathfrak{A}(z_2) = \iint (1 + p_2^2 + q_2^2)^{\frac{1}{2}}.$$

Thus $\mathfrak{A}(z_1)$ and $\mathfrak{A}(z_2)$ are both given by the classical integral formula. In this case however, as shown by McSHANE (see I.17), (6.11) implies that $z_2 - z_1$ is constant. Since $z_2 - z_1 = 0$ on C, it follows that $z_2 \equiv z_1$.

VI.20. McSHANE observed[1] that the preceding result yields the following generalization of the theorem of IV.10.

Let the continuous surface S be a solution of the least area problem for a JORDAN curve Γ^*. Suppose that every point P_0 of S is comprised in a portion S_0 of S which has a simply-covered orthogonal projection upon some plane. Then S is analytic (and is a minimal surface in the sense of differential geometry).

To prove this, consider any one of the portions S_0 described in the theorem, and suppose that S_0 has a simply covered xy-projection for instance. Denote by C any convex JORDAN curve, comprised with its interior in the projection of S_0. Denote by S_0^* the portion of S_0 which is projected into the JORDAN region bounded by C. Let $z = z_0^*(x, y)$ be the equation of S_0^* and denote by $\varphi_0^*(P)$ the function to which $z_0^*(x, y)$ reduces on C. Then $z_0^*(x, y)$ clearly solves the problem $\mathfrak{A}(z) = $ minimum for the convex JORDAN curve C and the continuous boundary function $\varphi_0^*(P)$. Since the solution of the problem is unique (see VI.19), $z_0^*(x, y)$ coincides with the analytic solution whose existence is secured by VI.18.[2] This proves the theorem.

In IV.10, S has been supposed to be a regular surface of class C'. This assumption implies that S has, in the small, a simply covered

---

[1] McSHANE: On a certain inequality of STEINER. Ann. of Math. Vol. 33 (1932) pp. 123—138.

[2] The existence theorem of IV.16 is not sufficiently general for the present application, on account of the three-point condition.

projection upon a properly chosen plane. The result of McShane shows that this property is sufficient in itself to secure the analytic character of S. Thus the analytic character of a solution of the least area problem depends upon *regularity in the topological sense*, rather than upon regularity with respect to differential coefficients. It would be interesting to decide if it is sufficient to suppose that S is, in the small, a Jordan surface (one-to-one and continuous image of the circular disc).

VI.21. Up to this time, we only considered surfaces of the type of the circular disc, bounded by a given Jordan curve. All the problems we discussed so far could be stated for surfaces of any given topological type, bounded by any number of given curves. We conclude this report by a brief review of certain results due to J. Douglas concerning problems of this generalized type[1].

Let there be given two non-intersecting Jordan curves Γ_1, Γ_2 in the xyz-space. Determine in the uv-plane a circular ring bounded by two circles C_1, C_2 with center at the origin and with radii R_1 and $R_2 > R_1$, and three functions $x(u, v)$, $y(u, v)$, $z(u, v)$ with the following properties[2].

1. $x(u, v)$, $y(u, v)$, $z(u, v)$ are harmonic for $R_1^2 < u^2 + v^2 < R_2^2$,
2. and satisfy there the equations $E = G$, $F = 0$, and
3. $x(u, v)$, $y(u, v)$, $z(u, v)$ are continuous in $R_1^2 \leq u^2 + v^2 \leq R_2^2$ and the equations $x = x(u, v)$, $y = y(u, v)$, $z = z(u, v)$ carry the circles C_1 and C_2 in a one-to-one and continuous way into the given Jordan curves Γ_1 and Γ_2 respectively.

VI.22. In case Γ_1, Γ_2 are both in the xy-plane, and for instance Γ_1 is enclosed by Γ_2, then the above problem requires to map the ring-shaped region bounded by Γ_1 and Γ_2 upon a circular ring in a one-to-one and continuous and in the interior conformal way, as is easily seen by a reasoning similar to the one used in V.19. As is well known in this special case, the circular ring cannot be given arbitrarily, the ratio of the radii of the (concentric) boundary circles being invariable under conformal mapping. Therefore, in the problem stated in VI.21, the ratio $q = R_1/R_2$ is to be considered as unknown.

VI.23. J. Douglas[3] treats the problem of VI.21 by generalizing his method for the one-contour case. Take a q satisfying $0 < q < 1$, and

---

[1] J. Douglas announced [see J. Douglas: Solution of the problem of Plateau. Trans. Amer. Math. Soc. Vol. 33 (1931) p. 264] that his methods are adequate for the solution of the most general problem (any number of boundary curves, minimal surface of any prescribed topological type).

[2] J. Douglas considers the n-dimensional problem (in the sense explained in V.19) and points out that the value of n does not make any difference. We restrict ourselves therefore to the case $n = 3$.

[3] The problem of Plateau for two contours. J. Math. Physics, Massachusetts Inst. Technol. Vol. 10 (1931) pp. 315—359.

two concentric circles C_1, C_2 with center at $u = 0$, $v = 0$ and with radii R_1, R_2 such that $q = R_1/R_2$. Denote by

$$T_1: x = \xi_1(\Theta), \quad y = \eta_1(\Theta), \quad z = \zeta_1(\Theta),$$
$$T_2: x = \xi_2(\Theta), \quad y = \eta_2(\Theta), \quad z = \zeta_2(\Theta),$$

monotonic transformations of the circles C_1, C_2 into Γ_1, Γ_2 respectively. Denote by $x_1(u, v)$, $y_1(u, v)$, $z_1(u, v)$ the harmonic functions, defined in $u^2 + v^2 < R_1^2$ and obtained by means of the POISSON integral formula, using $\xi_1(\Theta)$, $\eta_1(\Theta)$, $\zeta_1(\Theta)$ as boundary functions. Let $x_2(u, v)$, $y_2(u, v)$, $z_2(u, v)$ have a similar meaning for $u^2 + v^2 < R_2^2$ and $x_{12}(u, v)$, $y_{12}(u, v)$, $z_{12}(u, v)$ for $R_1^2 < u^2 + v^2 < R_2^2$. Put

$$A(T_1) = \tfrac{1}{2} \iint (E_1 + G_1), \quad u^2 + v^2 < R_1^2, \tag{6.14}$$

$$A(T_2) = \tfrac{1}{2} \iint (E_2 + G_2), \quad u^2 + v^2 < R_2^2, \tag{6.15}$$

$$A(T_1, T_2; q) = \tfrac{1}{2} \iint (E_{12} + G_{12}), \quad R_1^2 < u^2 + v^2 < R_2^2. \tag{6.16}$$

We use again E, F, G, with the proper subscripts, to denote the first fundamental quantities of a surface.

VI.24. Denote by $m(\Gamma_1)$, $m(\Gamma_2)$, $m(\Gamma_1, \Gamma_2)$ the greatest lower bounds of $A(T_1)$, $A(T_2)$, $A(T_1, T_2; q)$ for all possible monotonic transformations T_1, T_2 and for all values of q, $0 < q < 1$. Suppose these lower bounds are finite (the geometrical interpretation of this condition will be considered in VI.29).

It follows from the definitions that

$$m(\Gamma_1, \Gamma_2) \leqq m(\Gamma_1) + m(\Gamma_2).$$

J. DOUGLAS proves that *if*

$$m(\Gamma_1, \Gamma_2) < m(\Gamma_1) + m(\Gamma_2),$$

then the problem in VI.21 is solvable. The existence proof is similar to the one he developed for the one-contour case, although a great deal more involved. The proof is based on a detailed study of $A(T_1, T_2; q)$ in its dependence upon T_1, T_2, q. We are going to indicate the main steps.

VI.25. Develop the functions $\xi_1(\Theta)$, $\eta_1(\Theta)$, $\zeta_1(\Theta)$, which define the transformation T_1, into a FOURIER series. Denote by a_{1m}, b_{1m} the vectors whose components are the coefficients of $\cos m\Theta$, $\sin m\Theta$ respectively in the developments of $\xi_1(\Theta)$, $\eta_1(\Theta)$, $\zeta_1(\Theta)$. Let a_{2m}, b_{2m} have the same meaning with respect to T_2. Then DOUGLAS finds that

$$A(T_1, T_2; q) = \frac{4\pi\alpha^2}{-\log q}$$

$$+ \frac{\pi}{2} \sum_{m=1}^{\infty} m \frac{(1 + q^{2m})(a_{1m}^2 + b_{1m}^2 + a_{2m}^2 + b_{2m}^2) - 4q^m(a_{1m}a_{2m} + b_{1m}b_{2m})}{1 - q^{2m}},$$

where

$$\alpha = \tfrac{1}{2}(a_{10} - a_{20}).$$

It is necessary for the sequel to control $A(T_1, T_2; q)$ in the vicinity of $q = 0$ and $q = 1$. DOUGLAS finds, for q close to zero, the development

$$A(T_1, T_2; q) = A(T_1) + A(T_2) + \frac{4\pi\alpha^2}{-\log q} + q \times (\text{power series of } q). \quad (6.17)$$

For q close to 1, the following asymptotic expression is obtained:

$$A(T_1, T_2; q) \sim \frac{3\pi\alpha^2 + \frac{1}{2}\int_0^{2\pi} p(\Theta)\, d\Theta}{1 - q} \quad (6.18)$$

and

$$\frac{dA(T_1, T_2; q)}{dq} \sim \frac{3\pi\alpha^2 + \frac{1}{2}\int_0^{2\pi} p(\Theta)\, d\Theta}{(1 - q)^2}, \quad (6.19)$$

where

$$p(\Theta) = (\xi_1(\Theta) - \xi_2(\Theta))^2 + (\eta_1(\Theta) - \eta_2(\Theta))^2 + (\zeta_1(\Theta) - \zeta_2(\Theta))^2.$$

The numerators in (6.18) and (6.19) are clearly $\geq \pi d^2$, where d is the by assumption positive shortest distance of Γ_1 and Γ_2. DOUGLAS also observes that the approximations (6.18) and (6.19) are uniform, provided Γ_1 and Γ_2 are restricted to fixed bounded regions.

VI.26. By the definition of $m(\Gamma_1, \Gamma_2)$, there exists a sequence $(T_1^{(n)}, T_2^{(n)}; q^{(n)})$ such that $A(T_1^{(n)}, T_2^{(n)}; q^{(n)}) \to m(\Gamma_1, \Gamma_2)$. For a properly chosen subsequence, which we denote again by $(T_1^{(n)}, T_2^{(n)}; q^{(n)})$, $T_1^{(n)}$ and $T_2^{(n)}$ will converge toward transformations T_1, T_2 and $q^{(n)}$ will converge to a limit q. In the present case it is impossible to normalize the sequences $T_1^{(n)}$, $T_2^{(n)}$ as it has been done in the one-contour case by prescribing the images of several points, since the conformal maps of a circular ring upon itself are determined as soon as the image of a single boundary point is known. Consequently, the selection theorem of I.22 cannot be referred to directly. An easy discussion shows however that the limit transformation T_1 is either a monotonic transformation in the sense of I.21 or otherwise T_1 is *degenerate* in the sense that it carries the whole circle into a single point. The same holds for the limit transformation T_2.

The first thing to do is therefore to show that T_1, T_2 are non-degenerate and that q is different from 0 and from 1. Otherwise $(T_1, T_2; q)$ could not serve to solve the problem $A(T_1, T_2; q) = $ minimum. All these possibilities are excluded, as DOUGLAS finds, by the assumptions $m(\Gamma_1, \Gamma_2) < m(\Gamma_1) + m(\Gamma_2)$ and $m(\Gamma_1, \Gamma_2) < \infty$. It follows for instance from (6.17) for $q^{(n)} \to 0$ that

$$\lim A(T_1^{(n)}, T_2^{(n)}; q^{(n)}) \geq m(\Gamma_1) + m(\Gamma_2),$$

and hence, since $\lim A(T_1^{(n)}, T_2^{(n)}; q^{(n)}) = m(\Gamma_1, \Gamma_2)$,

that $m(\Gamma_1, \Gamma_2) \geq m(\Gamma_1) + m(\Gamma_2)$. For $q^{(n)} \to 1$ it follows from (6.18) that

$$A(T_1^{(n)}, T_2^{(n)}; q^{(n)}) \to +\infty.$$

VI.27. Thus we have a q with $0 < q < 1$, a circular ring $R_1^2 \leq u^2 + v^2 \leq R_2^2$ with $R_1/R_2 = q$, and monotonic transformations

$$T_1 : x = \xi_1(\Theta), \quad y = \eta_1(\Theta), \quad z = \zeta_1(\Theta),$$

$$T_2 : x = \xi_2(\Theta), \quad y = \eta_2(\Theta), \quad z = \zeta_2(\Theta)$$

of the boundary circles C_1, C_2 of the ring into the given JORDAN curves Γ_1, Γ_2, such that
$$A(T_1, T_2; q) = m(\Gamma_1, \Gamma_2).$$

From $m(\Gamma_1, \Gamma_2) < +\infty$ it follows again, by means of the expression (6.16) and, for instance, the lemma in V.15, that T_1, T_2 are both continuous.

VI.28. Denote by $x_{12}(u, v)$, $y_{12}(u, v)$, $z_{12}(u, v)$ the harmonic functions, defined in $R_1^2 \leq u^2 + v^2 \leq R_2^2$, which reduce to $\xi_1(\Theta)$, $\eta_1(\Theta)$, $\zeta_1(\Theta)$ on C_1 and to $\xi_2(\Theta)$, $\eta_2(\Theta)$, $\zeta_2(\Theta)$ on C_2. DOUGLAS shows that these functions satisfy the equations $E_{12} = G_{12}$, $F_{12} = 0$, by computing the first variation of the functional $A(T_1, T_2; q)$. The variations of T_1, T_2 are controlled by computations similar to those in V.18. In addition variations of q must also be considered in the present case. These are handled by means of the asymptotic expression (6.19). Thus it follows that $x_{12}(u, v)$, $y_{12}(u, v)$, $z_{12}(u, v)$ satisfy all the conditions of the problem, except possibly for condition 3. Now that $E_{12} = G_{12}$, $F_{12} = 0$ has already been shown, this last point can be taken care of easily. If the correspondence between the boundaries would not be one-to-one, then x_{12}, y_{12}, z_{12} would reduce to constants on a whole arc of the boundary of the circular ring, and consequently (cf. the end of V.10) they would reduce to constants identically. Then T_1, T_2 would necessarily both be degenerate, and this possibility has been excluded previously.

VI.29. There remains to discuss the *geometrical meaning of* the condition $m(\Gamma_1, \Gamma_2) < m(\Gamma_1) + m(\Gamma_2)$. Denote by $\mathfrak{P}_{12}^{(n)}$ a sequence of polyhedrons, of the topological type of the circular ring, such that the boundary polygons of $\mathfrak{P}_{12}^{(n)}$ converge, in the FRÉCHET sense, to Γ_1 and Γ_2 respectively. Consider $\underline{\lim} \mathfrak{A}(\mathfrak{P}_{12}^{(n)})$. Then DOUGLAS observes that the minimum of $\underline{\lim} \mathfrak{A}(\mathfrak{P}_{12}^{(n)})$, for all possible sequences $\mathfrak{P}_{12}^{(n)}$ with the properties required above, is equal to $m(\Gamma_1, \Gamma_2)$. Similarly, $m(\Gamma_1)$ and $m(\Gamma_2)$ are equal to the minimum areas of the curves Γ_1, Γ_2 in the LEBESGUE sense. The area of the minimal surface S whose existence has been proved in VI.23 to VI.28 is clearly equal to $m(\Gamma_1, \Gamma_2)$. Since obviously the area of any continuous surface, of the type of a circular ring, bounded by Γ_1 and Γ_2 is at least equal to $m(\Gamma_1, \Gamma_2)$, it follows that the area of the minimal surface S is a minimum with respect to all these surfaces. Thus it follows that *the condition $m(\Gamma_1, \Gamma_2) < m(\Gamma_1) + m(\Gamma_2)$ has the following geometrical meaning: the minimum of the areas of all continuous surfaces, of the type of the circular ring, bounded by Γ_1 and Γ_2 is less than the sum* *of the minima of the areas of all continuous surfaces, of the type of the circular disc, bounded by Γ_1 and Γ_2 separately.*

J. DOUGLAS quotes several cases in which the condition $m(\Gamma_1, \Gamma_2) < m(\Gamma_1) + m(\Gamma_2)$ is satisfied. An interesting case is that of two interlacing JORDAN curves. Another important case is that of two JORDAN curves Γ_1, Γ_2 in the same plane, such that Γ_2 for instance encloses Γ_1. In this case, the problem of VI.21 reduces to map the ring-shaped region between Γ_1 and Γ_2 in a one-to-one and continuous and in the interior conformal way upon a circular ring. J. DOUGLAS emphasizes the fact that his method yields a new proof for the existence of this map.

J. DOUGLAS observes that in order to obtain the geometrical meaning of $m(\Gamma_1, \Gamma_2)$, $m(\Gamma_1)$, $m(\Gamma_2)$ described above, it is necessary to use the existence of conformal maps of polyhedrons, in a similar way as in the one-contour case (see VI.5). He considers this again as a stop-gap and is planning to develop a method entirely independent of the theory of conformal mapping.

VI.30. In what precedes it has been supposed that $m(\Gamma_1, \Gamma_2)$ is finite. DOUGLAS generalizes his result in the following way. Denote by $e(\Gamma_1, \Gamma_2)$ the maximum of $\overline{\lim}[m(\Gamma_1^{(n)}) + m(\Gamma_2^{(n)}) - m(\Gamma_1^{(n)}, \Gamma_2^{(n)})]$ for all sequences $\Gamma_1^{(n)}, \Gamma_2^{(n)}$ such that $\Gamma_1^{(n)} \to \Gamma_1$, $\Gamma_2^{(n)} \to \Gamma_2$ and such that $m(\Gamma_1^{(n)}, \Gamma_2^{(n)})$ is finite. These conditions are certainly satisfied by polygons $\Gamma_1^{(n)}, \Gamma_2^{(n)}$, for instance. DOUGLAS obtains then the theorem that *the problem in* VI.21 *is solvable for* Γ_1, Γ_2 *if* $e(\Gamma_1, \Gamma_2) > 0$.

The method consists of solving the problem for a sequence $\Gamma_1^{(n)}, \Gamma_2^{(n)}$ such that $m(\Gamma_1^{(n)}) + m(\Gamma_2^{(n)}) - m(\Gamma_1^{(n)}, \Gamma_2^{(n)}) \to e(\Gamma_1, \Gamma_2)$ and passing then to the limit. The passage to the limit is substantially the same as that described in VI.26, and is controlled by the fact that the approximations (6.18) and (6.19) are uniform provided the curves involved are restricted to fixed finite regions of the space.

The condition $e(\Gamma_1, \Gamma_2) > 0$ is satisfied, as DOUGLAS observes, for *any two interlacing* JORDAN *curves.*

VI.31. Closely related to the two-contour case is the problem, considered by DOUGLAS, to *determine a minimal surface of the topological type of the* MÖBIUS *strip and bounded by a given* JORDAN *curve*[1]. The analytic formulation of the problem is as follows[2].

Given a JORDAN curve Γ, determine a q satisfying $0 < q < 1$ and three functions $x(u, v)$, $y(u, v)$, $z(u, v)$ with the following properties.

1. $x(u, v)$, $y(u, v)$, $z(u, v)$ are harmonic for $q^2 < u^2 + v^2 < 1$ and
2. satisfy there the equations $E = G$, $F = 0$.

---

[1] J. DOUGLAS: One-sided minimal surfaces with a given boundary. Trans. Amer. Math. Soc. Vol. 34 (1932) pp. 731—756.

[2] J. DOUGLAS considers the n-dimensional problem (in the sense explained in V.19) and points out that the value of n does not make any difference. We restrict ourselves therefore to the case $n = 3$.

3. $x(u, v)$, $y(u, v)$, $z(u, v)$ are continuous in $q^2 \leq u^2 + v^2 \leq 1$ and the equations $x = x(u, v)$, $y = y(u, v)$, $z = z(u, v)$ carry the circles $u^2 + v^2 = 1$ and $u^2 + v^2 = q^2$ in a one-to-one and continuous way into the given JORDAN curve Γ.

4. If $\xi(\Theta)$, $\eta(\Theta)$, $\zeta(\Theta)$ are the boundary values of $x(u, v)$, $y(u, v)$, $z(u, v)$ on $u^2 + v^2 = 1$, then the boundary values on $u^2 + v^2 = q^2$ are $\xi(\Theta + \pi)$, $\eta(\Theta + \pi)$, $\zeta(\Theta + \pi)$.

VI.32. Condition 4 implies that the equations $x = x(u, v)$, $y = y(u, v)$ $z = z(u, v)$ define a minimal surface of the type of the MÖBIUS strip. Indeed, if (u, v) is a point of $q^2 \leq u^2 + v^2 \leq 1$, then let (u^*, v^*) denote the point obtained by reflecting (u, v) on the circle $u^2 + v^2 = q$ and then rotating the resulting point around $(0, 0)$, by the angle π. The points (u, v), (u^*, v^*) will be called *elliptically inverse* to each other with respect to the circle $u^2 + v^2 = q$. Condition 4 expresses then that

$$x(u, v) = x(u^*, v^*), \quad y(u, v) = y(u^*, v^*), \quad z(u, v) = z(u^*, v^*) \quad (6.20)$$

on the boundary of the ring $q^2 \leq u^2 + v^2 \leq 1$. Consequently (6.20) also holds in the whole ring, since elliptic inversion transforms harmonic functions into harmonic functions, and since harmonic functions are determined by their boundary values. From (6.20) it follows that the equations $x = x(u, v)$, $y = y(u, v)$, $z = z(u, v)$ determine a minimal surface of the type of the MÖBIUS strip (every point of which is obtained twice).

VI.33. Denote by
$$T : x = \xi(\Theta), \quad y = \eta(\Theta), \quad z = \zeta(\Theta),$$
a monotonic transformation of $u = \cos\Theta$, $v = \sin\Theta$ into the given JORDAN curve Γ, and denote by T^* the transformation
$$T^* : x = \xi(\Theta + \pi), \quad y = \eta(\Theta + \pi), \quad z = \zeta(\Theta + \pi).$$

Denote by q a number satisfying $0 < q < 1$. Let $x(u, v)$, $y(u, v)$, $z(u, v)$ denote the harmonic functions, defined in $q^2 \leq u^2 + v^2 \leq 1$ by means of the POISSON integral formula, using as boundary functions $\xi(\Theta)$, $\eta(\Theta)$, $\zeta(\Theta)$ on $u^2 + v^2 = 1$ and $\xi(\Theta + \pi)$, $\eta(\Theta + \pi)$, $\zeta(\Theta + \pi)$ on $u^2 + v^2 = q^2$. Put
$$A(T; q) = \tfrac{1}{2} \cdot \tfrac{1}{2} \int\int (E + G),$$
the integration being extended over $q^2 < u^2 + v^2 < 1$. The method of DOUGLAS consists then in minimizing $A(T; q)$; the harmonic functions $x(u, v)$, $y(u, v)$, $z(u, v)$ corresponding to the minimizing set $(T; q)$ solve then the problem in VI.31. DOUGLAS observes that with the notations of VI.23 we can write
$$A(T; q) = \tfrac{1}{2} A(T, T^*; q).$$

Thus the present problem appears as a *limit case of the two-contour problem*. As a matter of fact, the treatment of the present problem follows step by step the treatment of the two-contour case, described in VI.23 to VI.30. We restrict ourselves therefore to state the result.

Denote by $\overline{m}(\Gamma)$ the greatest lower bound of $A(T;q)$ and let $m(\Gamma)$ have the same meaning with respect to Γ as in the one contour case. Then it is easily seen that $\overline{m}(\Gamma) \leqq m(\Gamma)$. *Whenever $\overline{m}(\Gamma) < m(\Gamma)$, the curve Γ bounds a minimal surface of the type of the* Möbius *strip.* The geometrical interpretation is similar to that in the two-contour case.

VI.34. The preceding condition presupposes that $\overline{m}(\Gamma)$ and $m(\Gamma)$ are finite. In complete analogy with the two-contour case, Douglas obtains the following generalization. Denote by $e(\Gamma)$ the maximum of $\overline{\lim}(m(\Gamma^{(n)}) - \overline{m}(\Gamma^{(n)}))$ for all sequences of Jordan curves $\Gamma^{(n)} \to \Gamma$ such that $m(\Gamma^{(n)})$ and $\overline{m}(\Gamma^{(n)})$ are finite. *Whenever $e(\Gamma) > 0$, Γ bounds a minimal surface of the type of the* Möbius *strip.*

VI.35. An application of the preceding criterion, considered by Douglas, is not quite convincing. Suppose the given Jordan curve Γ has upon the xy-plane a projection of the shape indicated in the figure. Denote by S the minimal surface of the type of the circular disc bounded by Γ. Then, according to Douglas[1], S should evidently have the general form of the Riemann surface of $(x + iy)^{\frac{1}{2}}$, that is to say S should evidently have a branch-point, and from this fact

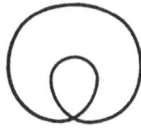

Fig. 1.

Douglas infers that Γ bounds a minimal surface of the type of the Möbius strip. Well then, it is obvious that if the xy-projection of Γ is prescribed as in the figure, the xz-projection can be chosen as a simply covered star-shaped curve, and then (see III.10) *none* of the minimal surfaces of the type of the circular disc bounded by Γ will have a branch-point. It should also be observed that if a Jordan curve Γ has an zy-projection as indicated in the figure, then Γ is certainly not knotted. Hence problem P_3 (see III.5) has a solution for Γ (see V.8). If a solution S of problem P_3 has a branch-point, then the order is at least 2 (see III.20). Every plane through a branch-point of S intersects therefore Γ in at least 6 distinct points (see III.8). Applying this remark to planes perpendicular to the xy-plane, it follows that whenever Γ has an xy-projection as indicated in the figure, then Γ bounds a minimal surface of the type of the circular disc which is free of branch-points.

These remarks show the interest and the necessity of a thorough investigation of the singularities of minimal surfaces in their dependence upon the shape of the boundary curve.

(Closed Dec. 15, 1932.)

---

[1] One-sided minimal surfaces with a given boundary. Trans. Amer. Math. Soc. Vol. 34 (1932) pp. 733, 739, 753.